**Mathematics Activities Handbook for Grades 5-12**

# Mathematics Activities Handbook for Grades 5 - 12

Michael C. Hynes and Douglas K. Brumbaugh

Parker Publishing Company, Inc.
West Nyack, New York

© 1976, by

PARKER PUBLISHING COMPANY, INC.

West Nyack, N.Y.

*All rights reserved. No part of this book may be reproduced in any form or by any means, without permission in writing from the publisher.*

Library of Congress Cataloging in Publication Data

Hynes, Michael C.
  Mathematics activities handbook for grades 5-12.

  1. Mathematics--Study and teaching.  I. Brumbaugh, Douglas K.,         joint author.  II. Title.
QA16.H95         510'.76        75-30963
ISBN 0-13-562280-8

Printed in the United States of America

To Cassie, Hilary, Mike, and Shawn.

# How to Use This Book for Dynamic Results in Teaching Mathematics

The activities in this sourcebook will help you to capitalize on the curiosity of students. You will discover that each activity is related to current classroom topics, can be used in career education, provides flexibility in classroom planning, and will motivate students to study and accomplish more in mathematics.

Career education, a concept which is broader than vocational education, is a process which provides students the opportunity to become acquainted with many vocations and avocations in our society. In mathematics classes, teachers must provide children with *relevant* experiences which will allow them to understand the importance of mathematical knowledge in various roles in society. Thus, many of the activities in this text have been designed to show the importance of mathematics in unexpected places in our society.

Teachers of mathematics are concerned about using many different modes of instruction, individualizing instruction, and using non-textbook oriented materials so that the general quality of their mathematics instruction may increase. Laboratory oriented activity lessons are compatible with all of those goals. The teacher may use activities to introduce topics, reinforce concepts, or to culminate a unit of study. The activities in this book are appropriate for large group, small group or individual instruction, depending upon the needs of the students.

In order to motivate students most effectively, we should recognize the need they have to study mathematics in a setting which is meaningful to them. Home economics, industrial arts, art, sports, games, puzzles, crafts, and many other areas of interest are used as themes for activities. Teachers may capitalize on the natural interest of students in such topics to make mathematics meaningful and fun to study.

Regardless of the way in which the activity is presented to children, the teacher who is using mathematics laboratories for the first time should begin gradually. Initially, the activity can be used with a small group of students. For many of those involved, both students and teachers, the laboratory approach will be a new and different experience. It is not safe to assume that because it is "different" the students will immediately like it and profit from it. Rather, a careful, planned introduction must be made in order to overcome the conditioned concept that mathematics classes are basically paper/pencil situations. Given this type of introduction, students and teachers alike will benefit from the activities in this book and proceed to a new, exciting and productive means of learning mathematics.

Activities to help you implement the laboratory approach can be found easily in this sourcebook in two ways. First, the Table of Contents provides you with a categorization of the activities according to major content area headings. Second, in deference to the present trend toward the use of behavioral objectives in mathematics instruction, the reader is provided with a list of behavorial objectives for mathematics. Each of the objectives is followed by a list of the numbers of the activities which can be used—either directly, or altered slightly in intent, materials, or level of sophistication—to achieve the objective.

The use of an objective key will perhaps be better understood by looking at an example. One of the generally accepted behaviors expected of mathematics is:

> The student will be able to write and solve mathematical equations.(3.02, 2.02, 2.03, 4.06)

Following the behavioral objectives are the numbers of the activities related to this objective. Some of the activities are designed specifically for the attainment of this objective. Other activities employ mathematical equations to attain another objective, or may be altered slightly so they may be used for achieving this objective.

Once the teacher identifies the desired major content area through either the Table of Contents or the Behavioral Objective List, the activity will be separated into two major sections—the teacher's section and teachers' comments to the students. The teacher's section contains a goal indicating the relevancy of the laboratory to the student, describes specific behaviorally stated objectives for the activity, indicates the appropriate grade levels for the activity, lists the necessary materials to set up the laboratory, provides instructions for teachers, and includes comments about the activity. The instructions for teachers provide information concerning the optimum number of students who can participate in the activity, the chronology of events completing the activity, the motivation needed for initiating the activity, etc. The "comments" section offers information such as safety tips and possibilities for altering the activities for use with different students, in different content areas, or with different materials.

The Teacher Comments to Students section is designed so the teacher may actually use the laboratory activity in the classroom directly from the book. For groups of students, the teacher will probably find it necessary to make duplicate student sheets for each student or groups of students. However, individual students may use the book itself to begin the activity since the teacher's section contains no information which should not be seen by the students.

## BEHAVIORAL OBJECTIVES KEYED TO ACTIVITIES

In each subject area the student will accomplish these goals.

**Algebra:**
1. write and solve mathematical equations. (1.09, 2.02, 2.03, 2.04, 2.05, 2.08, 2.09, 3.02, 4.06, 4.09, 5.08, 7.02, 8.07, 9.02, 9.04)
2. supply solutions to simultaneous equations. (2.05)
3. compute square roots. (2.05, 2.08)
4. factor polynomial expressions. (2.05, 2.06, 2.07, 2.08)
5. recall the vocabulary of algebra. (2.01, 2.05, 4.03, 8.08)

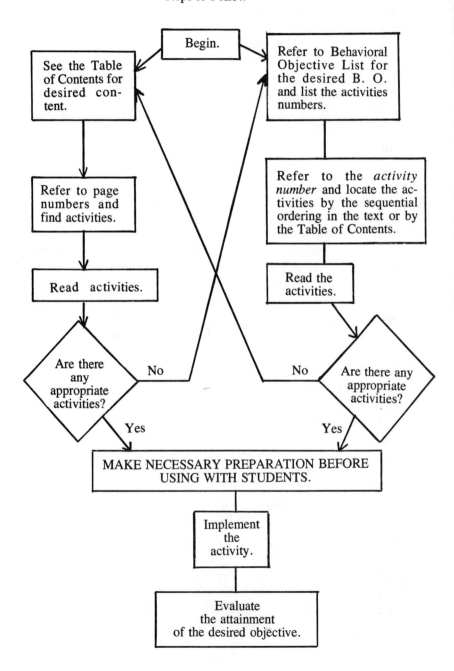

6. write the products of factors in algebraic form. (2.05, 2.07, 2.08)

**Career Education:**
   1. compute using a calculator. (2.03, 5.04, 5.08)
   2. apply mathematical concepts to career situations (2.03, 5.04, 5.05, 5.08, 6.10, 7.08, 8.03, 8.05, 9.02)
   3. apply mathematical concepts to consumer situations. (4.10, 5.01, 5.02, 5.03, 5.04, 5.05, 5.06, 5.08, 7.08, 8.04)
   4. apply mathematical concepts to ecological situations. (5.08, 8.06, 8.09)

**Graphing:**
   1. graph the solution(s) to equations and inequalities on the number line. (2.05, 8.01, 8.07, 8.08, 8.09, 8.10)
   2. graph the solution(s) to equations and inequalities on the Cartesian plane. (2.05, 3.08, 8.01, 8.07, 8.08, 8.09, 8.10)
   3. graph the solution to simultaneous equations on the Cartesian plane. (2.05)
   4. construct bar graphs, line graphs, circle graphs, pictographs, and histograms. (8.01, 8.02, 8.03, 8.04, 8.05, 8.06, 8.07, 8.09, 8.10)

**Integers:**
   1. indicate an understanding of the concept of integers. (4.08, 5.04)
   2. add integers. (2.05, 4.09, 5.04, 8.08, 8.10)
   3. subtract integers. (2.05, 4.09, 5.04, 8.08, 8.10)
   4. multiply integers. (4.08, 5.04, 8.08, 8.10)
   5. divide integers. (8.08, 8.10)

**Measurement:**
   1. measure linear distance. 1.01, 1.06, 1.08, 1.10, 2.03, 3.08, 4.03, 4.06, 4.11, 5.08, 6.08, 6.09, 7.03, 7.04, 7.05, 7.06, 8.07)
   2. measure and compute area. (1.02, 1.06, 2.08, 4.11, 5.08, 6.01, 6.09, 7.03, 8.01)
   3. measure and compute volume. (1.04, 1.06, 4.10, 5.08, 6.10, 8.10)
   4. measure liquids. (1.05, 1.06, 2.04, 4.10, 8.10)
   5. measure mass. (1.03, 1.06, 1.07, 4.10, 5.03, 8.07)
   6. measure angles. (6.03, 6.04, 6.07, 7.01, 7.02, 7.05)

7. measure time. (2.03, 2.05, 3.02, 4.06, 6.02, 8.01, 8.05, 8.07, 8.09)
8. compute rates. (2.03, 3.02, 4.06, 5.02, 5.06)
9. estimate measures. (1.02, 1.03, 1.04, 1.05, 1.06, 3.08, 4.10, 4.11, 5.03, 5.08, 7.02)

**Metric System:**
1. measure linear distance in the metric system. (1.01, 1.06, 1.08, 1.09, 1.10, 6.08, 6.09, 7.05, 8.07)
2. measure liquids in the metric system. (1.05, 1.06)
3. measure the mass of objects in the metric system. (1.03, 1.06, 1.07, 5.03, 8.07)
4. measure volume in the metric system. (1.04, 1.06, 6.10)
5. measure area in the metric system. (1.02, 1.06, 8.01)

**Geometry:**
1. make geometric constructions. (2.05, 6.02, 6.03, 6.04, 7.06, 7.10, 7.11)
2. engage in activities involving topological properties. (6.06, 7.01, 7.04, 7.05, 7.07)
3. construct three-dimensional geometric figures. (5.07, 6.02, 6.06, 6.07, 6.10, 7.05, 7.07)
4. make generalizations concerning the properties of two-dimensional figures. (2.05, 6.03, 6.04, 6.06, 7.01, 7.02, 7.04, 7.08, 8.08)
5. indicate a correct perception of spatial relations. (5.07, 6.06, 7.01, 7.03, 7.05, 7.07, 7.08)
6. identify geometric shapes and figures. (2.05, 2.06, 2.07, 2.08, 4.05, 5.07, 6.03, 6.04, 6.05, 6.07, 6.08, 6.09, 7.01, 7.02, 7.03, 7.06, 7.10, 7.11, 9.03, 9.05, 9.06, 9.07)
7. recognize congruent and/or similar shapes. (5.07, 6.05, 7.01, 7.02, 7.05, 7.06, 9.02, 9.03, 9.05, 9.06, 9.07)
8. identify the vocabulary of geometry. (6.07, 6.10, 7.08, 7.09, 7.10, 7.11, 9.02)

**Number Theory:**
1. recognize and identify prime and composite numbers. (9.05)
2. factor numbers into prime factors. (9.05)
3. extend given number patterns. (2.09, 4.01, 4.02, 6.02, 6.10, 7.02, 7.03, 7.04, 7.07, 9.01, 9.02, 9.03, 9.04, 9.05, 9.06, 9.07, 9.08)

*How to Use This Book for Dynamic Results* 15

    4. recognize properties of finite number systems. (9.03)
    5. form subsets of a given set. (6.02, 9.04)
    6. identify relations among sets. (3.04, 6.02, 9.01, 9.04)

**Probability and Statistics:**
    1. compute measures of central tendency. (3.07, 4.04, 4.06, 8.02, 8.05, 8.10)
    2. compute the probability of the occurrence of selected events. (3.01, 3.03, 3.04, 3.05, 3.06, 3.07, 3.09, 3.10)
    3. predict results of experiments based on collected data. (3.01, 3.02, 3.03, 3.04, 3.05, 3.06, 3.07, 3.08, 3.10, 8.01, 8.05)
    4. estimate probabilities. (3.01, 3.03, 3.04, 3.05, 3.06, 3.10)
    5. collect data for the purpose of statistical analysis. (3.02, 3.05, 3.07, 3.08, 3.10, 6.06, 8.02, 8.03, 8.04, 8.05, 8.06, 8.07, 8.09, 8.10, 9.01)
    6. define probability. (3.03, 3.04, 3.05, 3.06, 3.09, 3.10)

**Ratio, Per Cent:**
    1. make scale drawings. (2.05)
    2. compute per cent. (3.07, 4.05, 4.07, 8.02)
    3. compute ratio. (1.07, 1.08, 1.10, 3.07, 3.08, 4.03, 4.05, 4.06, 4.07, 5.01, 5.02, 5.06, 5.07, 7.06, 8.03, 8.06, 8.08, 8.10)

**Rational Numbers:**
    1. indicate an understanding of the concept of fractional numbers. (3.01, 3.03, 3.04, 3.05, 3.06, 3.07, 3.09, 3.10, 4.03, 4.04, 4.05, 4.06, 4.07 4.10)
    2. indicate an understanding of the concept of decimals. (4.05, 4.07)
    3. add fractional numbers. (4.10, 4.11, 5.01, 7.01)
    4. subtract fractional numbers. (4.10, 4.11, 5.01, 7.01)
    5. multiply fractional numbers (4.10, 4.11, 5.01, 5.03, 5.06, 5.08, 7.01)
    6. divide fractional numbers. (5.01, 5.02, 5.03, 5.06, 7.01)
    7. add decimals. (1.07, 1.09, 5.01, 5.04, 5.05)
    8. subtract decimals. (5.01, 5.04, 5.05)
    9. multiply decimals. (1.07, 1.09, 5.01, 5.02, 5.06, 5.08)
    10. divide decimals. (1.09, 3.07, 5.01, 5.02, 5.06)

**Application:**
    1. apply mathematical concepts to art forms. (4.04, 4.05, 4.06, 5.07, 8.04, 9.03)

2. apply mathematical concepts to homemaking. (3.01, 5.01, 5.02)
3. apply mathematical concepts to sports. (1.01, 1.06, 1.08, 1.09, 2.05, 4.06, 6.08, 6.09, 8.03, 8.07, 9.04)
4. apply mathematical logic to games and puzzles. (2.01, 2.02, 3.02, 5.04, 9.01)

**Whole Numbers:**
1. add whole numbers. (3.07, 4.04, 4.09, 5.01, 5.05, 8.05, 8.06, 9.01, 9.02, 9.06, 9.07, 9.08)
2. subtract whole numbers. (3.07, 4.04, 4.09, 5.01, 5.05)
3. multiply whole numbers. (4.01, 4.02, 4.09, 5.01, 5.03, 8.05, 9.05, 9.07)
4. divide whole numbers. (3.07, 4.04, 4.09, 5.01, 5.02, 5.03, 8.05, 9.05)

Michael C. Hynes
Douglas K. Brumbaugh

## ACKNOWLEDGMENTS

This book is designed for those who want to help children become actively involved in learning mathematics. Many have assisted us both directly and indirectly in the development of ourselves and the ideas projected here. We respectfully acknowledge the help of the following:

K-12 teachers
College teachers
College classmates
Jill Branding
Ann Jones
Tricia O'Neal
Students
Judy
Mary Ellen

# Table of Contents

**How to Use This Book for Dynamic Results in Teaching Mathematics** ............................................. 9

**1. Teaching the Metric System to Students with Motivating Activities** ............................................. 23

        Standing Metric Jump • 24
        Test of Metric Strength • 24
        Guess Your Mass • 26
        How Many Cubes? • 26
        Pour Metric • 27
        Metric Olympics • 28
        Weightwatcher's Delight • 29
        50-Meter Field Goal • 31
        Stay in Your Lane • 33
        Speedo • 35

**2. Activities for Teaching Algebra** ............................... 37

        Algecab • 38
        Equations with Exponents • 38
        Speed Trap • 40
        Pour It On • 42
        Bicycle Gymkhana • 44
        Squares • 46

**2. Activities for Teaching Algebra** *(Continued)*

    Square Sum • 49
    Color a Root • 51
    By "n—1" in "n" • 56

**3. Teaching Probability and Statistics** .......................... 59

    Alphabet Soup • 60
    Turtle Race • 61
    Paper Drop • 63
    The Rounder • 66
    Shake 'em Up • 68
    Paper Cup • 70
    Baseball—Who's Worth It? • 71
    Drop 'em • 73
    Draw • 75
    Color Wheel • 77

**4. How to Make Everyday Numbers Rational** .................. 81

    By Nines in Ten • 82
    Facto Beads • 84
    Rise and Run • 86
    Balloon Animals • 88
    100 % Mobile • 90
    Top Speed of a Ten-Speed • 92
    Pieces of % • 93
    "Lights; Camera; Action!" • 95
    Equations • 99
    No Chocolate Mess • 100
    How Much Material? • 101

**5. Mathematics Activities for Daily Living** ..................... 105

    The Best Price • 106
    Buy Right • 107
    The Street Value of Tobacco • 108
    Check Monopoly • 109
    Your Change, Mister! • 110
    What's the Difference? • 111

*Table of Contents*

**5. Mathematics Activities for Daily Living** *(Continued)*

    Mobile • 113
    Draggin' On • 115

**6. Activities That Teach Accuracy in Measurement..........119**

    Square Off • 120
    Anyday Calendar • 121
    Shape Up I • 124
    Shape Up II • 125
    SymMETRIC • 127
    Peel It! • 129
    Paper Protractor • 131
    Make a Kite! • 132
    Kite Pulling Contest • 138
    Boxed In • 139

**7. Simplified, Interesting Ways to Teach Geometry .........143**

    String It • 144
    Rip 'em Up • 146
    It's Obvious, Isn't It? • 150
    Dunebuggy Race • 152
    Soap Bubbles • 154
    Circular Shadows? • 156
    Tab It! • 158
    Styrofoam • 159
    Geocab • 161
    Fold 'em Up I • 161
    Fold 'em Up II • 164

**8. Developing Meaningful Data in Graphs .....................169**

    Milk Jugs • 169
    How Tall? • 171
    Tickets, Please! • 173
    Top Ten • 175
    Pulse Rate • 176
    Car Pool • 177
    Toboggan Run • 179

**8. Developing Meaningful Data in Graphs** *(Continued)*
>The Right Slant • 181
>Noise Pollution • 183
>Rain • 185

**9. Activities That Teach Number Theory to All Students ..189**
>Rat Maze • 190
>Sum Pages • 191
>Mod Art • 193
>Let's Get 'em • 196
>Prime Numbers • 198
>Square Numbers • 200
>Pentagonal Numbers • 202
>Magic Square Cubes • 204

# 1

# Teaching the Metric System to Students with Motivating Activities

Are your students apprehensive about learning a "new" system of measurement? Motivational activities which introduce students to the metric units of measure and reduce anxiety levels are presented in this chapter. Recreational activities, practical applications, and metric measure in sporting events are provided to appeal to the interests of many students.

The recreational activities are exploratory in nature and no previous knowledge of metric units is necessary to complete them successfully. Each of these activities may be used to introduce specific aspects of metric measure, or they may all be used together as a spectacular introduction to the study of the metric system.

Students are often concerned about how adoption of the metric system in the United States will affect their daily lives. For example, many students of general mathematics are eagerly anticipating being able to drive an automobile, but realize with anxiety that using the metric system will require some adjustments in driving habits. For example, although the time required to travel between two points will be the same (assuming no change), the rate will be much different when using the shorter kilometer in place of the mile (55 mph = 92 kmph). These students can be motivated through activities related to this topic which is interesting to them.

Sports activities may also be affected by the adoption of the metric system. How big will a football field become? How far will horses race? Baskets in basketball should be how high? What is longer—a 300-foot horse run or a 28-meter horse run? Use the activities related to sports in this chapter to motivate discussion and study of the metric system.

**TITLE:** **Standing Metric Jump (1.01)**

GOAL: The student will measure length in the metric system in a recreational activity.

OBJECTIVE: The student will measure the length of his/her standing long jump in meters and/or centimeters.

GRADES: 3, 4, 5, 6, 7, 8, General Mathematics.

MATERIALS: Three meter sticks, a record chart.

INSTRUCTIONS FOR TEACHERS: Place the three meter sticks on the floor end to end. Have the children place both feet at the beginning of the first meter stick with toes even with the end of the stick. The student is asked to jump in a line parallel to the meter sticks by taking off from both feet at once. The student records his/her distance on the metric chart.

COMMENTS: This is a noisy activity, and for safety purposes requires a 15-meter by 10-meter clear area. Thus, it is often an outdoor activity. There is no Teacher Comment to Students section for this activity. This activity may be included in a "Metric Olympics" by combining it with Activities 1.02, 1.04, and 1.05.

**TITLE:** **Test of Metric Strength (1.02)**

GOAL: The students will measure area in the metric system in a recreational activity.

OBJECTIVE: The student will count square centimeters and estimate parts of square centimeters to determine the area of an irregular area.

GRADES: 4, 5, 6, 7, 8, General Mathematics.

MATERIALS: Modeling clay, centimeter graph paper.

INSTRUCTIONS FOR Since this activity is noisy, the students will work best on the floor or outdoors on a smooth flat surface.

*Teaching the Metric System to Students with Motivating Activities* 25

TEACHERS:         Be certain that the students count the partial squares that are covered.
If a contest is desired, it is best to separate the results of the boys and girls to be fair.
COMMENTS:     This activity can be a part of the Metric Olympics to motivate the study of the metric system by using it with Activities 1.01, 1.03, 1.04, and 1.05.

TEACHER COMMENTS TO STUDENTS:

Roll the clay into a smooth spherical ball. Place the ball in the center of the graph paper. Hit the ball one time as hard as you can!

Use a pencil to draw an outline of your shape, and peel the clay from the paper.

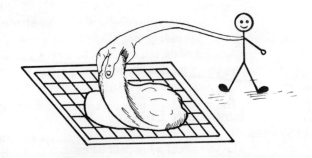

How many squares were covered by your flattened clay? Be sure that you count the partial squares!

Who is the strongest, metrically speaking?

| | |
|---|---|
| **TITLE:** | **Guess Your Mass (1.03)** |
| GOAL: | In a recreational lesson the students will have the opportunity to estimate and measure mass in metric units. |
| OBJECTIVES: | Given a kilogram mass, the student will estimate his/her weight in kilograms. |
| | The student will measure his/her weight on a metric scale. |
| | The students will graph the results of comparing the estimates of their weights with their actual metric weights. |
| GRADES: | 3, 4, 5, 6, 7, 8, General Mathematics. |
| MATERIALS: | One kilogram mass, one metric bathroom scale. |
| INSTRUCTIONS FOR TEACHERS: | Have each student lift the kilogram mass, and let each of them record his/her estimate of his/her weight. |
| | After all students have estimated their weights, let the students share their estimates to establish the range of estimates. |
| | Now, let all students weigh themselves (allow for some privacy here so that students are not embarrassed). Have the students share the differences between their estimates and their actual weights by using histograms. |
| | List the data on the chalkboard and ask the students to make a graph of their results. |
| COMMENTS: | This activity has no Teacher Comments to Students section. |
| | This activity may become a part of the Metric Olympics by having one team try to guess the total weight of the other team. The team with the closest guess wins. For the Metric Olympics in this activity, use Activities 1.01, 1.02, 1.04, and 1.05. |
| **TITLE:** | **How Many Cubes? (1.04)** |
| GOAL: | The volume of a box will be estimated in a situation similar to guessing how many beans are in a jar. |
| OBJECTIVE: | The student will estimate the number of centimeter cubes needed to fill a given container. |
| GRADES: | 4, 5, 6, 7, 8, General Mathematics. |

MATERIALS: A handful of centimeter cubes (unit Cuisenaire Rods or Centicubes), a box from a grocery item, a box for the guesses.

INSTRUCTIONS FOR TEACHERS: Place the box for which the volume will be estimated on a table in the mathematics interest center. Put the centimeter cubes near the box in a pile, but do not let the students "measure" with them. Have each student place his/her estimate in a container.

At the end of a designated period of time, allow the student to fill the box with centimeter cubes to determine the volume. Select the closest estimate, announce the winner and award a prize.

COMMENTS: There is no Teacher Comments to Student section for this activity.

This activity could be included in a Metric Olympics for those students who are not highly motivated by physical activities. If this is done, use it with Activities 1.01, 1.02, 1.03, and 1.05.

TITLE: **Pour Metric (1.05)**

GOAL: This activity is designed as a competitive task which will motivate and introduce liquid measure in the metric system.

OBJECTIVE: The student will pour water among three beakers to arrive at a given volume in one beaker.

MATERIALS: Three plastic containers for each group.

GRADES: 7, 8, Algebra I.

INSTRUCTIONS FOR TEACHERS: Form teams in the class. Groups of four or five make the best teams for this activity. Explain the rules of the contest, and stress the fact that the first answer with the least number of steps is the winner.

COMMENTS: If this activity is used with Activities 1.01, 1.02, 1.03, and 1.04, it may become a part of the Metric Olympics.

If an alternate problem is desired, try using 600 ml., 500 ml., and 300 ml. beakers to achieve 400 ml. in one beaker. Of course, there are many other examples that you may wish to use!

If you do not have enough beakers for the whole

class, any containers with the 800 ml., 500 ml., and 300 ml. marks will suffice.

TEACHER COMMENTS TO STUDENTS:

You have been given three containers, and these containers will hold 800 milliliters, 500 milliliters, and 300 milliliters respectively.

Fill the 800 ml. container. Now, your task is to pour the water into the other containers so that one container has 400 ml. in it. When you pour the water, you must either empty the container from which you are pouring or fill the container which you are filling. In other words, you may not estimate!

Record your answers like this:

(800 ml., 0, 0)
(0, 500 ml., 300 ml.)

The answer with the fewest number of steps wins.

| TITLE: | **Metric Olympics (1.06)** |
|---|---|
| GOAL: | Students who are fearful of learning a new system of measurement need to be introduced to the metric system in a non-threatening manner. This activity will allow students to become familiar with the metric system of linear, square, cubic, liquid, and mass measurement. |
| OBJECTIVE: | The student will participate in a recreational activity involving the metric system. |
| MATERIALS: | See Activities 1.01, 1.02, 1.03, 1.04, and 1.05. |
| GRADES: | 7, 8, General Mathematics. |
| INSTRUCTIONS FOR | Make up a chart to keep score in the "Olympics," and make a paper first prize for the winning team. |

*Teaching the Metric System to Students with Motivating Activities*

| | |
|---|---|
| TEACHERS: | Divide the class into five teams and announce the rules:<br>1. Every team must enter every event,<br>2. Every team member must participate in a least one event, and<br>3. The scoring will be 6 points for first place in an event, 4 points for second place, 3 points for third place, 2 points for fourth, and 1 point for fifth.<br><br>The order of events will be:<br><br>1. Standing Metric Jump (two team members and use successive jumps).<br>2. Test of Metric Strength (two team members with one hit each).<br>3. Guess Your Mass (team event).<br>4. How Many Cubes (two team members consult for one answer).<br>5. Pour Metric (team event).<br><br>In announcing the events for the Metric Olympics, the most effective procedure is to announce them separately. Announce that the first event will be the Standing Metric Jump so that teams may chose contestants who are appropriately dressed for this event. The remainder of the events are more fun if the teams must choose participants before the event. This creates suspense and maintains interest. Since no other event is strenuous or requires casual dress, there is no danger in this procedure. |
| COMMENTS: | Refer to Activities 1.01, 1.02, 1.03, 1.04, and 1.05 for specific materials and instructions for the events.<br><br>If a Metric Olympics is held you must plan approximately two hours of class time for the entire extravaganza.<br><br>There is no Teacher Comments to Student section for this activity. |
| **TITLE:** | **Weightwatcher's Delight (1.07)** |
| GOAL: | This activity utilizes the adolescent's interest in his weight—whether it be the football player trying to add muscular mass or the cheerleader trying to |

|                | keep slim—to motivate work in ratios and the metric system. |
|----------------|---|
| OBJECTIVES:    | The student will write equivalent ratios. |
|                | The student will make a kilogram dial for a bathroom scale. |
|                | The student will multiply and/or add decimals. |
| GRADES:        | 5, 6, 7, 8, General Mathematics. |
| MATERIALS:     | An old bathroom scale, paper, tape, one kilogram mass. |
| INSTRUCTIONS FOR TEACHERS: | Before this activity is begun, allow the students to estimate their weights in kilograms. |
| COMMENTS:      | This activity is based on the concept of ratios, but the same results could be obtained using angle measure. The central angle indicating a kilogram on the dial for the scale could be measured. Once this angle, approximately 1.6°, has been determined, all the other kilogram marks can be put on the dial by measuring central angles or intercepting equal arcs. |

### TEACHER COMMENTS TO STUDENTS:

How much do you weigh?

Did you answer this question by stating your weight in pounds? If so, do you know your weight in a metric unit?

Let's try to figure out how to measure a person's weight in kilograms. Take a kilogram mass and put it on a set of bathroom scales. How many pounds does a kilogram weigh?

The ratio of kilograms to pounds is approximately $\frac{1}{2.2}$.

Remove the dial from the bathroom scales and mark it in kilograms rather than pounds. To do this, first make a table of equivalencies for kilograms.

| Lbs. | 2.2 | | | | | | | | | |
|------|-----|---|---|---|---|---|---|---|---|----|
| Klg. | 1 | 2 | 3 | 4 | 5 | 6 | 7 | 8 | 9 | 10 |

Now, cut a circular piece of paper the same size as the dial from the bathroom scale, and place it on top of the dial. Using your table of equivalent values, mark the dial with kilograms and put the scale back together.

How much do you weigh in kilograms?

## TITLE: 50-Meter Field Goal (1.08)

GOAL: Recently, professional football owners, players and officials have been discussing the role of the field goal in professional football since the field goal, as a means of scoring points, is dominating the other ways to score. The controversy can be used to motivate the consideration of a metric football field.

OBJECTIVES: The student will measure a football field in the metric system.

The student will propose alterations to a football field which will decrease the potency of the field goal as a scoring device.

MATERIALS: Metric tape measure.

GRADES: 7, 8, General Mathematics.

INSTRUCTIONS FOR TEACHERS: If possible let the entire class visit a football field. Remember, many girls and boys have never had the first-hand experience of standing on a football field to get a feeling of its size. Kicking a field goal looks easy when you are in the stands, but looking at the goal posts from the 30- or 40-yard line gives a person a deeper respect for the field goal kicker's ability.

A discussion of the field goal controversy before this activity is begun would be helpful. Many boys realize that the hash marks on the field were moved in through a recent rules change. This means that the angle of the field goal has been narrowed; thus, the field goal is more likely to be successful. Also, many football "experts" feel that if defensive alignments such as the zone pass defense were outlawed the importance of the field goal would be reduced because more touchdowns would be scored. Finally, the goal posts, in professional football, are on the goal line. Some people would like to see the goal posts moved to the back of the end zone to increase the distance that field goals would be kicked from any given yard marker. All of this discussion will naturally lead you to interject the necessity of someday changing the 100-yard football field to metric measures.

COMMENTS: You may wish to have students make scale drawings of their metric football fields to support their arguments concerning how the effectiveness of the field goal can be reduced.

TEACHER COMMENTS TO STUDENTS:

1. What is the length of a football field using metric measure?
   _____
2. What is the width of a football field using metric measure?
   _____
3. 10 yards = _____ meters.
4. The depth of the end zone from the goal line to the back of the end zone is _____ meters.
5. Let's visit the football field again. Measure the length and width of the area inside the stadium. Would it hold a football field which measures 50 meters wide and 100 meters long plus 10-meter end zones on both ends?

6. If your area would not allow changing to a 100-meter field, which would be best, a 90-meter field with 10-meter end zones or an 80-meter field with 10-meter end zones? _____ Why? _____

7. If you could change the rules of football, how would you alter the rules to reduce the importance of the field goal? Remember, changing to the metric system should play a part in your answer. _____

| | |
|---|---|
| **TITLE:** | **Stay in Your Lane (1.09)** |
| GOAL: | The events in international track and field have been described or measured in the metric system for many years. This fact and the fairness of races offer many opportunities to explore mathematics. |
| OBJECTIVES: | The student will measure the distance around a track in various lanes. |
| | The student will determine the lane stagger necessary to establish a fair race for races where the runners must remain in their respective lanes. |
| | The student will compute and/or measure lane staggers on a 400-meter track. |
| GRADES: | 7, 8, General Mathematics, Geometry. |
| MATERIALS: | 100-foot measuring tape, a 50-meter tape. |
| INSTRUCTIONS FOR TEACHERS: | The questions in this activity are developmental and the students should be encouraged to answer each question before going on. |
| COMMENTS: | This activity can be accomplished entirely within the classroom with the use of scale drawings, but taking the students to a track increases the motivation for mathematics. |

TEACHER COMMENTS TO STUDENTS:

If we were going to have a one-lap race around the track, in which lane would you want to run? (See following illustration.)

Yes, Lane 1 is the shortest, but everyone cannot run in the

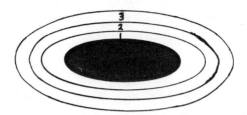

same lane. To overcome this difficulty, track officials stagger the starting lines in successive lanes.

A racetrack is 440 yards around, and the lanes are 42 inches wide. The track is composed of two straight sections and two half-circle curves. A runner who runs one lap around the track in the middle of Lane 1 travels 440 yards.

Make a scale drawing of a track of this type, and determine how far a person in Lane 4 runs in one lap.

How much farther does a runner in Lane 4 have to run than one in Lane 1?

How much should the starting line in Lanes 2, 3, and 4 be advanced so that a 440-yard race is fair?

Fill in the chart shown in Figure 1-1.

|  | 220 yds. | 440 yds. | 880 yds. | Mile |
|---|---|---|---|---|
| Lane |  |  |  |  |
| Stagger |  |  |  |  |

**FIGURE 1-1**

What if we were entering an international race where the metric system is used? What would the lane staggers be for races on a 400-meter oval track with one-meter-wide lanes? (See Figure 1-2.)

|  | 200 m. | 400 m. | 800 m. | 1500 m. |
|---|---|---|---|---|
| Lane |  |  |  |  |
| Stagger |  |  |  |  |

**FIGURE 1-2**

*Teaching the Metric System to Students with Motivating Activities*     35

**TITLE:**     **Speedo (1.10)**

GOAL: When the metric system of measure arrives in the United States, people will need to convert their car speedometers from English to metric. These converters are available now through commercial distributors, but it is a relatively simple process to make one and much mathematics will be involved.

OBJECTIVES: The student will establish equivalent metric expression (in kmph) for the multiples of ten found on speedometers in the United States.

The student will make a metric speedometer for his family car.

The student will install his metric speedometer in the family car.

MATERIALS: Clear contact paper, permanent marking pencil, access to a car.

GRADES: 6, 7, 8, 9, General Mathematics.

INSTRUCTIONS FOR TEACHERS: This activity requires some coordination with home since the student will be altering the family car. As the student rounds values, he should be cautioned to round up when possible to insure his being within the legal speed limit when driving.

COMMENTS: This activity can be done one of two basic ways. Either the metric speed for each multiple of ten on the common speedometer can be used or a new mark can be established for every ten kilometers per hour. Each method has an advantage: the first in that the marks are already established, thus making them easy to locate, and the second in that the new numerals can be located between the existing ones. Although in this activity the student will place the contact paper on the glass or plastic cover of the speedometer, some students may elect to remove the speedometer and cover the existing numerals.

TEACHER COMMENTS TO STUDENTS:

As the number of distance signs expressed in kilometers increases, signs expressing speeds in terms of kilometers per hour will probably begin appearing and then it would be helpful to know your rate of speed in kilometers per hour. Rather than buying a stick-on cover for your car's speedometer, you can make one with clear contact paper and a permanent marker.

Make a pattern of the glass or plastic cover of the speedometer in your car. On this pattern indicate the location of the 10, 20, 30 mile per hour marks on the speedometer. Your speedometer is shown in Figure 1-3.

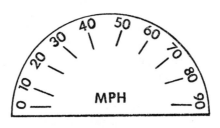

**FIGURE 1-3**

You now need to determine the 10 miles per hour equivalent in kilometers per hour. A book of measures can be used to get the fact that 1 mile = 1.6093 kilometers. Rounding to the nearest hundredth, 1 mile = 1.61 kilometers. Thus, 10 miles = 16.1 kilometers, and position A in Figure 1-4 would be 16.1 or 16. The first mark at the lower left is still zero. To get the metric equivalent of 20 miles per hour (either 10 mph times 2 or 10 mph plus 10 mph), either multiply 1.61 kmph by 20 or add 16.1 kmp to 16.1 kmph. Proceed in a like manner until you have the metric equivalent expression for the English values on your speedometer.

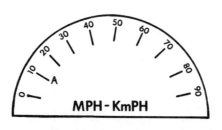

**FIGURE 1-4**

# 2

# Activities for Teaching Algebra

There is much more to algebra than the memorization of rules and the manipulation of symbols. Algebra can be brought to life through geometric interpretation of concepts, games related to skills and terminology, and application to the students' everyday life.

Geometric interpretation of concepts such as differences of squares, squares of sums, and square root provides students a semiconcrete representation of these concepts which are often difficult when presented in the abstract mode. Many students of algebra begin floundering when only abstract presentations of concepts are made, but creative teachers strive to illustrate concepts in as many ways as possible. Some of the activities in this chapter will be models for illustrating many algebraic concepts more concretely.

Games are often employed by teachers of pre-algebra mathematics to create interest and to maintain skills. Games can also be used in algebra class to introduce topics, to provide a culminating activity for a unit of study, and to enrich the curriculum as well as to create interest and to maintain skills.

Meaningful applications of algebra for students are necessary to maintain students' awareness of the relevance of studying algebra. Thus, concerned teachers of algebra seek situations from the students' everyday life. There are many activities which can be developed from the students' interests; "Speed Trap," "Pour It On," and "Bicycle Gymkana" are examples of such activities which the resourceful teacher of algebra can use to stimulate the development of many more meaningful activities for algebra.

| | |
|---|---|
| **TITLE:** | **Algecab (2.01)** |
| GOAL: | Students enjoy activities where they are given arrays of letters in which they are to find words. Often these arrays are of a general nature, but the idea can easily be adapted to productive classroom use. |
| OBJECTIVE: | The student will, given an array of letters, locate and loop Algebra I vocabulary words. |
| MATERIALS: | Student Algecab I sheet. (See Figure 2-1.) |
| INSTRUCTIONS FOR TEACHERS: | Students need to know the words are written horizontally, vertically and diagonally. Furthermore, some of the words are in reverse order. |
| COMMENTS: | This activity can be used in a contest manner to see who can get the most words, the largest number of the same word, a specified number of words, the most words within a designated time period, a particular word the fastest, etc. |
| | Perhaps you would like to make your own Algecab. Determine the size of the array, write in the appropriate words and fill in the empty cells randomly. Be sure to check carefully for obscene words. |

| | |
|---|---|
| **TITLE:** | **Equations with Exponents (2.02)** |
| GOAL: | The student will recognize exponential functions on whole numbers in a puzzle. |
| OBJECTIVE: | The student will write the symbols for the four fundamental operations, the equality sign, parenthesis, and a star or asterisk to indicate exponents to form equations with whole numbers. |
| MATERIALS: | Puzzle sheet. |
| GRADES: | 8, General Mathematics, Algebra. |
| INSTRUCTIONS FOR TEACHERS: | This activity must be introduced by showing a new way to write exponents. Instead of $3^2 = 9$, the students will write $3*2 = 9$. |
| | Motivation for the game may be provided by beginning a contest to determine who can form the greatest number of equations involving exponents. Students of high ability may wish to make their own puzzles. |

# Activities for Teaching Algebra

```
Q U R O S I V I D J S E Q U I V A L E N T T X
V A R I A T I O N M R E U J B L X I U T J P I
V S B I N A R Y E O C X A O M I I N M U C A R
L S T S U B S E T N A P D T I W O E V E C R T
G O M R O C E A T O E O R S T E M A T R A A A
J C N A T L E N U M G R A L I A G R E E R B M
P I R F U O U N C I T I O N O R H C A L T O D
Y A F O P S E T N A I I D E N T I T Y R E L I
T T R N T E E R E L N V G A L P J N I Q S A 5
H I R A Y D U N A V L N T H R I C G T Y I U C
A V S J B M O U E L A T A O R O O T N E A O R
G J E Q U I A R X R T L C I O N M N S J N S I
O T N T N E S L P F R A U S M E P C T I L C M
R Y E U L E O H O R L A W E Y D L R N A N O I
E F U N C T I O N D N M T H E D E U I M E A N
A A M D L N G A E I J R C A L O X G L M A M A
N C I S A D U R N A I R R A T I O N A C E U N
S T J L E S H E T C O O R D O N A T E D N L T
M O A O P E N D E S R O R E L A T I O N B T E
P R O P O R T I O N A L I G C D W I R M T I R
A T N E L J E S L D E T E R M I N A N T S P E
L A M G R A P H R A S N D E S H L A W N E L M
C S A D N J A M O D A H I E M P T Y S E T E I
```

**FIGURE 2-1**

COMMENTS: Similar puzzles can be made for equations in the sets of whole numbers (Activity 4.09).

TEACHER COMMENTS TO STUDENTS:

Equations appear in this grid horizontally, vertically, and diagonally (Figure 2-2). Insert +, −, ×, ÷, =, parenthesis, and * to form as many equations as you possibly can. * is used to show

|   |   |   |   |   |   |   |   |   |
|---|---|---|---|---|---|---|---|---|
| 8 | * | 2 | = | 64 | 1 | 2 | 3 | 7 | 1 |
| 6 |   | 4 |   | 3 | 1 | 27 | 1 | 4 | 0 |
| 5 |   | 2 |   | 32 | 1 | 27 | 1 | 4 | 0 |
| 9 |   | 1 |   | 2 | 8 | 25 | 27 | 8 | 31 |
| 4 |   | 3 |   | 64 | 6 | 2 | 3 | 36 | 6 |
| 7 |   | 16 |   | 4 | 8 | 2 | 1 | 49 | 16 |
| 1 |   | 18 |   | 36 | 7 | 2 | 2 | 3 | 27 |
| 64 |   | 9 |   | 4 | 8 | 32 | 1 | 25 | 1 |

**FIGURE 2-2**

exponents. That is, $8^2 = 64$ is shown in the puzzle as 8*2 = 64. Be sure that each of your equations involves exponents!!

**TITLE:**      Speed Trap[1] (2.03)

GOAL: The students will develop and use a speed trap. The principle of the trap is similar to those used by the Highway Patrol.

OBJECTIVE: The students will measure both distance and time.

GRADE: 7 - 12. The sophistication of the equations, precision of measurement and discussion related to the establishment of the actual speed trap permit great fluctuation in the application level.

MATERIALS: Tape measure, stopwatch, pencil, paper, and a road with suitable traffic.

INSTRUCTIONS FOR TEACHERS:
1. Caution students as to the necessary care to be exercised when near the highway.
2. Calculation of the time required to travel the distance at the legal speed will provide a quick means of determining speeders.
3. Present each group with the student directions that follow.
4. Encourage students to construct "traps" of varying lengths and in areas that have different speed limits.
5. You may wish to have outside speakers to demonstrate how vascar and radar work.

---

Brumbaugh, D.K., and Hynes, M.C. "Math Lab Activities: Speed Trap," *School Science and Mathematics* (74), February 1974, p. 75. Initial idea posed by Merle B. Grady.

COMMENTS: As an independent project for one student, or as a challenge to the better students, ask them how to make a speed trap that can be manned by one student. The drawing below illustrates only one of several ways that this may be accomplished.

TEACHER COMMENTS TO STUDENTS:

   Here is how to establish a speed trap similar to those used by the police! First decide how long the speed trap will be. It should be at least 100 yards long. Now select a straight stretch of road and measure the desired distance. Mark the beginning and end of the trap well.
   To operate the trap, a person should be at each end ready to signal when a car enters and leaves the trap. One person needs a stopwatch to time the car. When considering the car's entrance and departure from the speed trap, be sure to use the same part of the car. For example, use the front bumper.
   Time several cars through your speed trap and record the times in a chart.
   In your computations, you will calculate the average speed of the car in the trap. The time required for each car to pass through the measured distance is measured. By using the formula "distance equals rate times time ($d=rt$)," the average rate of speed can be calculated. If the measured distance is 100 yards, you must convert the distance to miles and the time to hours.

For example:

$$4 \text{ seconds} = 4 \text{ seconds} \times \frac{1 \text{ minute}}{60 \text{ seconds}} \times \frac{1 \text{ hour}}{60 \text{ minutes}} = \frac{4}{3600} \text{hours}.$$

**TITLE:** **Pour It On (2.04)**

GOAL: We, as teachers, often state there are a variety of ways to do problems, and yet seldom provide opportunities involving manipulative materials which encourage a variety of solutions. In writing equations, one way of doing something often becomes so appealing that, when encountering a certain type of situation, we devote little or no thought to it and the same old procedure is followed. This activity, involving pouring water or sand, is designed to show a way to generate a variety of equations for one problem.

OBJECTIVE: The student will write equations to describe physical phenomena.

GRADE: Pre-Algebra, Algebra I, or Algebra II.

MATERIALS: Four graduated cylinders, water or sand, assignment cards such as the following:

| Problem | Given Amounts | | | Desired Results |
|---|---|---|---|---|
| | A | B | C | R |
| 1 | 28 | 2 | 5 | 19 |
| 2 | 7 | 24 | 3 | 13 |
| 3 | 19 | 61 | 5 | 28 |
| 4 | 20 | 29 | 24 | 13 |
| 5 | 23 | 49 | 3 | 6 |
| 6 | 8 | 27 | 4 | 15 |

INSTRUCTIONS FOR TEACHERS: Tape can be used to mark the given amounts on the graduated cylinders. The student is, through some combination of the given amounts in containers A, B, and C, to arrive at the amount listed under the desired result and then write an equation for the way he arrived at the result. Comparison of equations by different students will show that there are many ways of obtaining the desired solution from the given amounts. Actually pouring materials should encourage getting the most direct solution.

COMMENTS: It is possible that a student would follow a pattern rather than look for another solution; this is not always the best thing to do. For instance, examples 2, 3, and 4 are all solvable with the equation $B - 2A + C = R$, a usable pattern in example 5. This pattern does work in number 5, but a much easier one would be $2C = R$.

One advantage of providing a set of examples all solvable by the same equation is that the slower student has a "built-in" help that permits him to do all the problems correctly.

By pouring the materials into a fourth graduated cylinder, a ready check is provided.

TEACHER COMMENTS TO STUDENTS:

Often there are many different ways of getting an answer to a problem. In this activity you are to try to find a number of ways of getting a given result and write an equation for each way that works.

Your activity kit will contain four graduated cylinders and some problem cards. You will be directed to use something that can be poured easily, probably either sand or water. The problem card will contain a series of exercises, one of which might be:

| Give Amounts | | | Desired Result |
|---|---|---|---|
| A | B | C | R |
| 24 | 2 | 5 | 17 |

Some solutions to this particular problem are:

$$A - B - C = R$$
$$5C - 4B = R$$
$$6B + C = R$$
$$B + 3C = R$$

At times a pattern will work for more than one exercise in a set, as, for example, in the following exercises:

| Exercise | Given Amount | | | Desired Result |
|---|---|---|---|---|
| | A | B | C | R |
| 1 | 4 | 3 | 2 | 0 |
| 2 | 8 | 5 | 10 | 6 |
| 3 | 11 | 4 | 6 | 13 |

All these can be done by using the equation $2A - 3B + \frac{C}{2} = R$. But number three can also be done with the equation $A + C - B = R$, which in this case is easier than the pattern.

### TITLE: Bicycle Gymkhana[2] (2.05)

GOAL: Automobile gymkhanas are common in many areas. The ideas can readily be converted to bicycles, and the students can be encouraged to establish their own route and then have other groups "run" the course.

OBJECTIVE: The student will, using Cartesian coordinates, slope, geometric description, and magnetic compass readings, establish a gymkhana on an unfamiliar course for other students.

MATERIALS: Multi-speed bicycles, several clipboards, several stopwatches, maps of the route.

GRADES: 7 - 12.

INSTRUCTIONS FOR TEACHERS: In gymkhanas, the objective is to travel at legal or designated speeds along a route which is camouflaged in some manner. The camouflage here will be limited to Cartesian coordinates and magnetic compass readings (or turn right or left so many degrees).

For example, suppose the course illustrated at the top of the facing page is scaled so that one unit equals 100 meters. The directions might be: Face north at (0,0); turn counterclockwise to a line having slope (−1/2) and go to (−2,1); turn clockwise to a line having slope (−3/2); etc.

---

[2]Brumbaugh, D. K., and Hynes, M. C. "Math Lab Activities: Bicycle Gymkhana," *School Science and Mathematics* (74), April 1974, p. 341.

*Activities for Teaching Algebra* 45

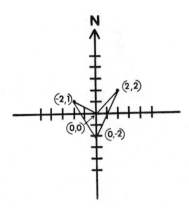

                  Suppose that, in this example, the students traveled in a given bicycle gear only and at a given pedaling rate (which could change for different legs of the course). The student should cover the distance in a prescribed number of seconds and each second over or under the prescribed time results in a penalty point. The person with the fewest penalty points wins.

COMMENTS:    This experiment could be conducted either on school grounds or inside the building (if the students learn to take a "standard" pace and then proceed at so many paces per time unit). In either instance, it might be advantageous to establish "false checkpoints" if the students can see checkpoints or other students going through the course. For assistance in establishing speed, see the experiment from this series on the top speed of a ten-speed bicycle.

TEACHER COMMENTS TO STUDENTS:

    Sports car clubs frequently have gymkhanas which are races involving driving at legal speeds.

    Your teacher will describe the area in which you can hold the gymkhanas and divide your class into groups. Each group will devise a course using Cartesian coordinates and directions. For example, suppose the following, simple routes were determined:

    Starting at (0,0), head north to checkpoint (0,2). Turn clockwise 90° and go to checkpoint (3,2). Head south to checkpoint

(3,1); head east to checkpoint (4,1). Turn 90° to the right to checkpoint (4,0). Turn counterclockwise 270° and proceed to checkpoint (0,0), the finish.

You will need to establish the desired speed (select a gear on a bicycle at x pedals per minute or establish a standard stride length with y strides per minute—see top speed of a ten speed) for each leg of the course and calculate the distance between checkpoints.

Establish a gymkhana course for the area designated by your teacher, being certain to keep your course a secret from the other groups. Make a scale drawing of the course, including directions necessary to get from each checkpoint to the next. Calculate the time required between checkpoints and actually run the course as a cross check. Prepare a set of instructions for your gymkhana course (no maps) and have another group try to go through it. Each group should run the course of every other group to determine who is the gymkhana champion of the class.

- What scale will you use to draw the map?
- Use geometric shapes in the directions!
- Use combinations of compass directions and headings in your directions!
- Slope of lines can be included in your directions!
- Descriptions of elevation may be used in the directions!

## TITLE: Squares (2.06)

GOAL: It is frequently difficult for a student to rationalize why some of the results of factoring come out as they do. This activity will provide the students with a pictorial description of the difference of two squares.

OBJECTIVES: The student will write the factorization of the difference of two squares.

GRADES: Algebra I, Algebra II.

MATERIALS: Student work sheet consisting of a series of instructions, and a pair of scissors for each student.

INSTRUCTIONS FOR TEACHERS: Provide each student with the materials and let them proceed at their own rate. You should have additional related activities for those who finish first.

*Activities for Teaching Algebra* 47

COMMENTS: This type of process can be used in other situations. Two of the more common ones are: $(a + b)^2 = a^2 + 2ab + b^2$ and finding the square root of a given value. Similar activities can be developed.

TEACHER COMMENTS TO STUDENTS:

Sometimes algebraic situations can be more clearly understood by using pictures or models. This activity is one such case. You will need scissors and a set of instructions, both of which will be provided by your teacher.

Get, or make, a square piece of paper. A square can be made from a rectangle by folding the paper so one of the short sides is made to coincide with one of its adjacent long sides. Cut the rectangular piece off, leaving a right triangle (Figure 2-3).

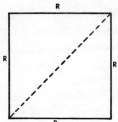

**FIGURE 2-3**

Opened out, the triangle paper gives a square of side R (Figure 2-4).

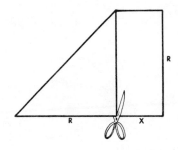

**FIGURE 2-4**

Note that each leg of the triangle is of length R. Make a fold parallel to either leg of the triangle, the fold being a distance S from either acute angle (Figure 2-5).

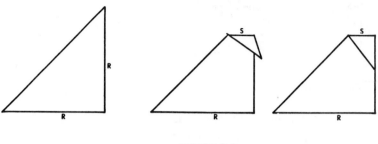

**FIGURE 2-5**

The distance from the fold to the right angle is _____.

Cut or tear along the fold, separating the smaller triangle. This leaves two congruent trapezoids folded one on top of the other. They are joined by the fold which was once the hypotenuse of the isosceles right triangle of leg length R. (See Figure 2-6.)

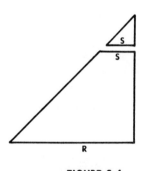

**FIGURE 2-6**

Opening the trapezoids, you get the area, which can be represented by $R^2 - S^2$. Cut along the fold, and flipping one trapezoid over, put the two trapezoids together to form a rectangle (Figure 2-7).

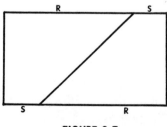

**FIGURE 2-7**

The length of the new rectangle is _____.
The width of the new rectangle is _____.
Therefore, the area of the new rectangle is expressed by _____ × _____. But this area is also $R^2 - S^2$. So, $R^2 - S^2 =$ _____ × _____.

Thus, in general, it can be said that: The _____ of two squares is the product of _____.

## TITLE: Square Sum (2.07)

GOAL: Students are frequently confused as to the origin of some terms in products of binomials. In particular, they are frequently bothered by the 2xy in $(x + y)^2$.

OBJECTIVE: The student will use concrete materials to show $(x + y)^2 = x^2 + 2xy = y^2$.

GRADE: Algebra I.

MATERIALS: Paper, pencil, straight-edge.

INSTRUCTIONS FOR TEACHERS: The initial discussion should review that in a square all side measures and angle measures are equal. Furthermore, $x^2$ is an algebraic description of the geometric representation of a square with side length x. This discussion should be extended to a square with side length $x + y$. At this point, the student should draw a square, label the dimensions, compute the area of each part, and add the resultant areas.

COMMENTS: This activity could be related to our activity, Squares. It can also be extended to the product of the sum and difference of two values by having the student select values for x and y. When the selected values are used to depict the situation, the drawing can be cut apart and a rearranging of the pieces will yield $x^2 - y^2$.

TEACHER COMMENTS TO STUDENTS:

You will need a square piece of paper. Express the length of the side of the square as $x + y$, where x is not necessarily equal to y. Mark a point on each of two adjacent sides a distance of x from the common vertex. (See Figure 2-8.) The remaining distance on each side will be y.

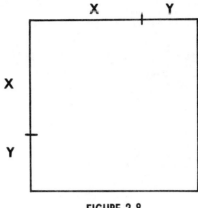

**FIGURE 2-8**

Draw lines parallel to the sides of the square through the points designating x's distance from the vertex (Figure 2-9).

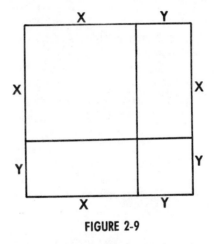

**FIGURE 2-9**

The square should now be divided into four parts: a large square, a small square and two rectangles. Notice that both rectangles are the same size.

Label both dimensions (length and width) for each of the four parts of the square.

- The area of the large square part is represented by_____.
- The area of the small square part is represented by_____.

*Activities for Teaching Algebra*     51

- The area of the rectangle is represented by _____.
- The area of the whole figure then would be given by _____.

**TITLE:**     **Color a Root (2.08)**

GOAL: Many square roots are irrational numbers, and students must use an abstract algorithm to compute them. Since computation on the abstract level with no concrete model is often meaningless and difficult, this activity is intended to provide a more concrete model for determining square roots.

OBJECTIVES: The student will determine approximate square roots using approximation techniques.

The student will solve second degree equations.

MATERIALS: White paper, Magic Markers, water colors or crayons, rulers.

GRADE: Algebra I.

INSTRUCTIONS FOR TEACHERS: Intuitively the square root can be illustrated as the length of the side of a square of a given area. For example, if 16 square units is given area, the measure of the length of a side of a square which would have the area of 16 square units would be four units.

More abstractly, the square root is described as one of two equal factors of a number. With this description, perfect squares are readily handled but non-perfect squares present problems. In the situation involving non-perfect squares, students sometimes erroneously conclude that there is no square root.

However, many square roots are irrational numbers and the algorithm usually used for computation of square roots is troublesome for many students. Even if a student can apply the algorithm, he has no idea how it works in many cases.

Suppose you have a square of area 746 square units. Find the length of each side. This square contains the smaller known square of area 625 square units which has side length of 25 units.

Taking out the known area of 625 square units, the area of the original square still unaccounted for is

746 − 625 or 121 square units. Algebraically,
$$746 = (25 + x)^2$$
$$= 625 + 50x + x^2$$
and $x(50 + x) = 121$.

x can be estimated by dividing the 121 by 50, giving approximately 2. If $x = 2$, then $x(50 + x)$ becomes $2(50 + 2) = 104$ which is less than 121. Therefore, a square larger than the estimated one of a 25 unit side is contained in the original square having area of 746 square units. See Figures 2-10 and 2-11.

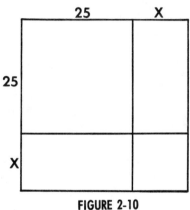

**FIGURE 2-10**

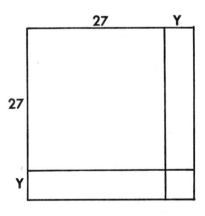

**FIGURE 2-11**

*Activities for Teaching Algebra* 53

$746 - 729 = 17$ and the remaining area can be expressed in y in $(27 + y)^2 = 746$. As before the remaining area would yield an approximation of y by dividing 17 by 54, giving y approximately as 0.3. So y $(54+y)$ becomes $0.3 (54+0.3) = 16.29$ which is less than 17. The process can be repeated again and again, thus making the result as accurate as desired.

The activity in the Teacher Comments to Students section accomplishes this method of approximation by coloring the estimated squares as the process continues.

COMMENTS: Please note the similarity between this method of approximation and the algorithm for square root.

```
        2
     √746.0000                1st estimate 20
       400.0000               746 = (x + 20)²
  4__  346.0000               746 = x² + 40x + 400
                              346 = x(40 + x)


        27
     √746.0000                2nd estimate 27
       400.0000               746 = (y + 27)²
  47   346.0000               746 = y² + 54y + 729
       329.0000               15 = y(54 + y)
  54    15.0000
```

## TEACHER COMMENTS TO STUDENTS:

Put away your calculator! Put away your slide rule!
What is the square root of 746?
If you said someplace between 25 and 30, you are beginning to color a root.
Let's look at a square with an area of 746 square units (Figure 2-12). What are the measures of the lengths of the sides of this square?

**746 sq. units**

**FIGURE 2-12**

We know that this square contains a square 25 units on each side.

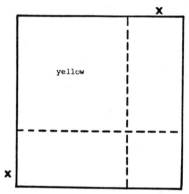

**FIGURE 2-13**

Color the 625 square units yellow, and label the amount left over on each side as x (Figure 2-13).

How much area is outside the smaller square of 625 square units?

Look at this!  $746 = (25 + x)(25 + x)$
$746 = 625 + 50x + x^2$
$746 = 625 = 50x + x^2$
$121 = 50x + x^2$
$121 = x(x + 50)$

x is approximately equal to _____.

Did you estimate x was 2? If you did, let's look at what has happened. First, the estimated square inside has been increased to an area of $27 \times 27$ or 729 square units.

## Activities for Teaching Algebra

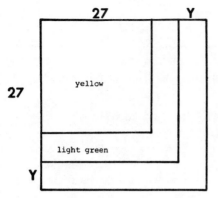

Color the new part of the estimated square light green (Figure 2-14).

**FIGURE 2-14**

Algebraically, we have:
$$746 = (27 + y)(27 + y)$$
$$746 = 729 + 54y + y^2$$
$$746 - 729 = 54y + y^2$$
$$17 = 54y + y^2$$
$$17 = y(54 + y)$$
  y is approximately equal to _____.

.3 is a good approximation for y, so (see Figure 2-15)...

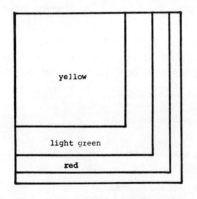

$$746 = (27.3 + z)(27.3 + z)$$
$$746 = 745.29 + 54.6z + z^2$$
$$746 - 745.29 = \boxed{\phantom{xxxx}}$$
$$\boxed{\phantom{xxxx}} = \boxed{\phantom{xxxx}}$$
$$\boxed{\phantom{xxxx}} = \boxed{\phantom{xx}}\,\boxed{\phantom{xx}}$$

**FIGURE 2-15**

Estimate a value for z. How near to 746 is $(27.3 + z)^2$ using your value for z?

Is this value, 27.3 plus your value for z, the square root of 746? _____ Why? _____

Using the process, find the square root of 585 to the nearest tenth. Show a colored picture for each step of the process.

| | |
|---|---|
| **TITLE:** | **By "n−1" in "n" (2.09)** |
| GOAL: | Students are often intrigued by unique or "gimmicky" approaches to topics. Multiplication by "n−1" in base "n" on the bead computer is a quick attention getter. |
| OBJECTIVE: | Using an apparatus made from wire and beads, the student will be able to multiply any single digit by "n−1" in base "n". |
| MATERIALS: | Coat hanger wire, five red beads and five blue beads, file, pliers. |
| GRADES: | 6, 7, 8, General Mathematics. |
| INSTRUCTIONS FOR TEACHERS: | See Facto Beads, Activity 4.02, to discover how to make the calculator. Although these instructions specify ten beads, the number could be altered to "n" by adding or taking away. For "n" less than ten, it is easier to move the "10−n" beads to the right and use only the left part of the calculator.
Be sure to check the rules before the student gives them to another person to try.
This pattern can be extended in a manner similar to that discussed in By Nines in Ten, Activity 4.01. By Nines in Ten should be done before this activity. |

TEACHER COMMENTS TO STUDENTS:

The calculator used in Facto Beads is to be used here.

You will be multiplying single digits by "n−1" in base "n". Place n beads at the left end of the calculator and 10−n at the right. The 10−n beads will not be used. Note that if the base is greater than 10, beads will have to be added.

Suppose you were multiplying by 4 in base 5. You would have five beads at the left end of the calculator and five more, not to be used, at the right end. To multiply 2 by 4, count to the second bead from the left. The beads to the left of that second bead represent the fives digit of the product and the beads to the right of the second bead give the ones digit of the product. That is, in base five, 4 × 2 would be shown in Figure 2-16.

**FIGURE 2-16**

In base five, 4 × 3 would be shown in Figure 2-17.

**FIGURE 2-17**

In base seven, 6 × 3 would be shown in Figure 2-18.

**FIGURE 2-18**

In base eight, 5 × 7 would be shown in Figure 2-19.

**FIGURE 2-19**

Develop a series of rules that tell the steps that must be used when multiplying by "n−1" in base "n" on this calculator. After you have checked to be certain those rules work, give your rules and the calculator to someone and see if he can multiply by "n− 1" in base n using the calculator.

# 3

# Teaching Probability and Statistics

The topics of probability and statistics are usually presented to students through laboratory activities. Experienced teachers are well aware of the typical activities for these topics involving cards, dice, coins, spinners, population bar graphs, and circle graphs for budgets. Even though many activities for probability and statistics are available, the concerned teacher is searching for activities which appeal to students of both sexes, varied backgrounds, different capabilities and diverse interests.

Often boys and girls are motivated to study mathematics through different activities. In this chapter, the teacher will find activities related to sports and recreation which may appeal more to boys, and activities relating to cooking and grooming which will appeal more to girls. Once these activities are used with children, the creative teacher will be able to devise similar activities which relate to the interests of students of both sexes.

Since in all probability many of the traditional activities involve the use of gambling devices, some teachers avoid teaching probability concepts and parents sometimes do not approve of the study of probability. However, there are many applications of probability which relate to areas other than gambling. Simple devices made with paper can be used in place of dice, or paper cups may be flipped rather than coins. Thus, children of different backgrounds may study the concepts of probability in less controversial situations.

The importance of studying concepts of probability and statistics involves other topics such as computational skills with whole numbers. However, the classroom teacher who is attempting to provide for individual differences often uses probability and statistical topics as enrichment lessons. Since individualizing instruction usually requires that some students work independently, some of the activities in this chapter are designed so that students of almost any ability can complete them with a minimum of supervision.

It's important to remember that activities in probability and statistics require that the students use other mathematical concepts and skills. Thus, most of these activities are keyed to many related objectives. The task of adapting activities from a level which requires computation with rational numbers expressed as decimals to one which requires computation with whole numbers or vice versa is left to the teacher.

**TITLE:** **Alphabet Soup (3.01)**

GOAL: Probability concepts occur in the strangest places. Even eating lunch may provide the opportunity if the soupe du jour is alphabet soup.

OBJECTIVE: The student will compute the probability of a given letter or set of letters appearing in a soup spoon when eating alphabet soup.

MATERIALS: Alphabet soup, bowls (various sized), spoons, cookie sheet, pan, large pot.

GRADES: 7, 8, General Mathematics.

INSTRUCTIONS FOR TEACHERS: This activity is most effective when 10 or 12 cans of soup are heated in a large pot in the cafeteria. By using a large number of cans, you are somewhat assured that you have a random sample of letters in the soup.

COMMENTS: This activity can be extended for the very able students. What is the probability that in three spoonfuls of soup, each containing one letter, the three letters make a word? This type of question requires that the students know something about the compatibility of letters in forming words. For example, if a "q" were one of the letters, would a "u" be needed to form a word? Is there more than one three-letter word containing a "q"?

TEACHER COMMENTS TO STUDENTS:

Get a can of alphabet soup and open the can without completely removing the lid. Pour the liquid into a pan, and dump the letters and vegetables carefully onto a cookie sheet.

How many letters are there in one can of soup?_____ Assume that all of your cans of soup have approximately the same number of letters. Now, pour the contents of all the cans of soup into the large pot along with the previously drained-off liquid. How many cans did you use?_____ Approximately how many letters are in all the cans?_____

Good! Now, let's try some experiments! If you use a teaspoon, what is the probability that you can predict that a certain letter is on the spoon in 20 trials? Try your guess? (See Figure 3-1.)

| Trial | 1 | 2 | 3 | 4 | 5 | 6 | 7 | 8 | 9 | 10 | 11 | 12 | 13 | 14 | 15 | 16 | 17 | 18 | 19 | 20 |
|---|---|---|---|---|---|---|---|---|---|---|---|---|---|---|---|---|---|---|---|---|
| Get an a? | | | | | | | | | | | | | | | | | | | | |

**FIGURE 3-1**

In how many trials did at least one "a" appear in the spoon?

What do you think of the probability of selecting six "a's" in one spoonful?

How many letters would have to fit on a spoon before you would predict an "a" on the spoon more than 90% of the time? Justify your answer.

In your spoon of six letters what is the probability of forming a word of three or more letters?

**TITLE:**       **Turtle Race (3.02)**

GOAL:        $D = rt$ is a familiar formula, but it is not very motivational unless the students have an opportunity to actually do some measuring.

OBJECTIVES:        The student will compute the speed of his turtle in

|              | miles per hour and/or kilometers per hour in a turtle race. |
|--------------|---|
|              | The student will use the data collected in the turtle race to predict distances and times given his turtle's speed. |
| MATERIALS:   | Several turtles, a circular ring made of masking tape or rope. |
| GRADES:      | 5, 6, 7, 8, General Mathematics. |
| DIRECTIONS FOR TEACHERS: | Turtle races will generate a great deal of excitement among the students. This enthusiasm can be used to promote computation of distances, rates and time. Let the winner of the race be not the fastest turtle but the most accurate computations based on the speed of the turtle. |
| COMMENTS:    | This is a classroom activity which can be extended by using Activity 2.03, Speed Trap. Food could be placed at the finish line to entice the turtle to move faster. This can become rather noisy. |

TEACHER COMMENTS TO STUDENTS:

Choose a turtle! Place it in the center of the "race track" so that it faces the outside of the circle (Figure 3-2).

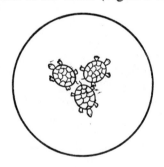

**FIGURE 3-2**

Get your stopwatches ready! Let the turtles go!
How long did it take your turtle to get to the finish line?
The distance from the starting line is_____.
Now compute the speed of the turtle._____

The speed of the turtle is helpful in predicting other results. If the turtle proceeds at the same rate for one hour, how far will it travel? _____ Try some more; fill in the chart shown in Figure 3-3.

|              | Turtle A | Turtle B | Turtle C |
|--------------|----------|----------|----------|
| Turtle speed |          |          |          |
| Time         |          |          |          |
| Distance     |          |          |          |

**FIGURE 3-3**

**TITLE:** **Paper Drop (3.03)**

GOAL: Common classroom materials can be used to generate probability activities. In this activity a piece of paper or an index card is used in two ways to allow the students to experiment with probability.

OBJECTIVES: The student will, given selected objects, establish the probability of the objects landing in a certain way when dropped from a specified distance.

The student will discuss biasing probability experiments.

GRADES: 7, 8, General Mathematics.

MATERIALS: ½ sheet of paper, tape, pencil, student activity sheet.

INSTRUCTIONS FOR TEACHERS: Every student in the class can take part in this activity. The materials are readily available and inexpensive, and little noise is caused by the dropping of paper.

Before initiating the second part of this activity using a taped piece of paper, talk with the students about biasing the results of chance occurrences in probability experiments.

COMMENTS: You may wish to use this activity in conjunction with Activity 3.06, Paper Cup. Both these activities can be used in a "probability day" to provide the students with many opportunities to explore the concept of probability instruction in a fun way.

TEACHER COMMENTS TO STUDENTS:

Using the rectangle shown in Figure 3-4 as a pattern, cut a rectangle out of a piece of notebook paper and fold it on the dotted

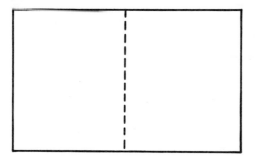

**FIGURE 3-4**

line so that the edges at the fold form approximately a 90 degree angle (an angle formed by two planes is called a dihedral angle). Drop the paper from a height of about two meters. There are three feasible outcomes for how the paper will land, each of which is shown in Figure 3-5.

**FIGURE 3-5**

If the paper lands on edge, the crease line will be perpendicular to the surface. Drop the paper 20 times and record the results (Figure 3-6).

| Outcome | Tally | Frequency | Part of Total Trials |
|---------|-------|-----------|----------------------|
| Up      |       |           |                      |
| Down    |       |           |                      |
| On Edge |       |           |                      |

**FIGURE 3-6**

# Teaching Probability and Statistics

Using the results from dropping the paper 20 times, predict the probable outcome for each possibility if the paper were dropped ten times, and record your predictions. Check these predictions by dropping the paper ten more times, using the same procudure as before, and record your results (Figure 3-7). Were your predictions exactly right? Were you close? Compare your results with those of other students. Are they similar?

| Outcome | Prediction | Tally | Frequency | Part of Total Trials |
|---------|------------|-------|-----------|----------------------|
| Up      |            |       |           |                      |
| Down    |            |       |           |                      |
| On Edge |            |       |           |                      |

**FIGURE 3-7**

Perhaps you had some difficulty in keeping the paper folded at a right angle in this activity. This could be fixed by placing a piece of tape on the paper as shown in Figure 3-8. The center of the tape should be halfway down each edge and the paper should be held at approximately a right angle when the tape is affixed.

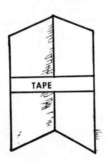

**FIGURE 3-8**

Do you think that the tape will change the probability of the paper landing a certain way? If it does, then, in terms of the original untaped paper, the results will be biased. After taping your paper, drop it 20 times, and record the results (Figure 3-9).

| Outcome | Tally | Frequency | Part of Total Trials |
|---------|-------|-----------|----------------------|
| Up      |       |           |                      |
| Down    |       |           |                      |
| On Edge |       |           |                      |

**FIGURE 3-9**

| | |
|---|---|
| **TITLE:** | **The Rounder (3.04)** |
| GOAL: | Probabilities are not always easily predicted. This activity presents the student with a situation that is basically not predictable. |
| OBJECTIVE: | The student will establish probabilities for an unknown situation. |
| GRADES: | 7, 8, General Mathematics. |
| MATERIALS: | Flat washer, paper, pencil, and game board (Figure 3-10). |

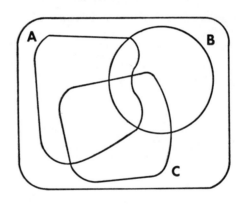

**FIGURE 3-10**

| | |
|---|---|
| INSTRUCTIONS FOR TEACHERS: | The Rounder game board must have boundaries to aid in determining when the washer is off the playing surface. If the washer slides off the surface, the throw isn't counted. When the washer lands on a line, the looped surface which is covered by the |

*Teaching Probability and Statistics*

|                | greatest part of the washer is the part of the board counted. |
|----------------|---|
| COMMENTS:      | Point values can be awarded for the different parts of the surface. |

TEACHER COMMENTS TO STUDENTS:

You are to determine the probability of a washer landing on a particular part of the Rounder game board by actually throwing the washer underhand at the surface one hundred times. You should be at least two meters from the game surface to make your throws. Each throw should be aimed at the dot in the center of the surface.

As the throws are made, each result should be recorded as a tally mark in the appropriate row of the tally column (Figure 3-11), the row being determined by the section of the surface the washer lands on. After one hundred throws are made, the tallies are recorded in the number of hits column and then that total is divided by 100 for the probability for the washer landing on each part of the game surface.

| Part of board hit | Tally | No. of Hits | Probability |
|---|---|---|---|
| A but not B∪C |  |  |  |
| B but not A∪C |  |  |  |
| C but not A∪B |  |  |  |
| A∩B but not C |  |  |  |
| A∩C but not B |  |  |  |
| B∩C but not A |  |  |  |
| A∩B∩C |  |  |  |
| (outside all loops) |  |  |  |

**FIGURE 3-11**

1. How do your probabilities compare to your classmates' findings?
2. Do you think your probabilities would be the same if you repeated this experiment? Why or why not?
3. What is the average of all the probabilities for each area found by all your classmates?
4. Do you think this average is a better estimate of the probability? Why or why not?

**TITLE:** **Shake 'em Up (3.05)**

GOAL: One desirable goal in education is to make students more aware of the objects around them which can be used in mathematics. This activity uses an egg carton and 12 markers. Using these common materials, this activity provides experience in predicting and testing probabilities.

OBJECTIVE: The student will predict and test a probable outcome.

GRADE: 7, 8, General Mathematics.

MATERIALS: Egg carton, three red markers, four white markers, five blue markers, pencil and chart (Figure 3-12).

INSTRUCTIONS FOR TEACHERS: Selection of the markers is significant in that they cannot be too large to prohibit their moving about as the egg carton is shaken. Care should be taken to insure that the students hold their cartons by the edge that opens to prevent the markers from flying out while shaking the carton. Finally, when the carton is opened (flat side down) the student should not look at the markers as one is drawn. If necessary, the carton could be held over the head during the draws. After noting the color, the marker is to be replaced.

COMMENTS: Variations of this activity could include drawing without replacement and considering how the probability will change for each color as the individual color numbers and the totals change. Consideration could also be given to drawing two markers at a time with replacements and the probability of drawing two of the same color for each color as well as drawing two different colors for all possibilities without paying attention to order. Better students could be challenged by a situation involving drawing two markers where order is significant. Note that this activity can easily be done by individuals.

TEACHER COMMENTS TO STUDENTS:

Of the 12 markers given you for this activity, what fractional part of the total number of markers is red? White? Blue? What is the sum of these fractions?

The probability of an event is defined as the total possible outcomes (total number of markers in this instance) divided into the number of favorable outcomes. For example, the probability of drawing a red marker would be 3/12 or 1/4.

You are to place the markers in the egg carton and shake them up, remove one of the markers, record the color in the chart provided, and replace the marker. Before actually performing the activity, you are to predict how many of each color will be drawn and record those estimates in the chart. If it would be expected that three reds would be drawn out of 12 tries, then three times three or nine reds would be expected if 36 draws are to be made, since 36 is 12 time three.

As you shake the carton, hold it upside down and by the edge that opens to prevent the markers from flying out. When you are removing the markers, hold the carton above your head so you cannot see the markers. If you can see the markers as you draw you may bias the results which means you might change the outcomes to be closer to what you predict.

How close did your experimental results come to the predicted results? What do you suppose the results would be if the number of drawings were increased more and more? Why did you make the supposition you did?

| COLOR OF MARKER | ESTIMATED RESULTS IN 36 DRAWS | OBSERVED RESULTS IN 36 DRAWS |
|---|---|---|
| red | | |
| white | | |
| blue | | |

**FIGURE 3-12**

| | |
|---|---|
| **TITLE:** | **Paper Cup (3.06)** |
| GOAL: | Frequently students are asked to predict outcomes of situations in which the results are "known" because of the objects used. In this activity the students will be confronted with a situation in which the expected outcome is an unknown and must be established. |
| OBJECTIVE: | The student will, given selected objects, establish the probability of the objects landing a certain way by actually dropping the objects. |
| GRADES: | 7, 8, General Mathematics. |
| MATERIALS: | Paper cup, pencil, student activity sheet. |
| INSTRUCTIONS FOR TEACHERS: | The students need to be encouraged to actually drop the objects to establish the expected outcomes. This process will provide them with a means of solving a situation in which the outcome cannot be predicted; that is, they should learn that the information can be gathered and used as a basis for future work. |
| COMMENTS: | Similar activities can be done with common household articles such as toothpaste tube lids, silverware, unbreakable containers, etc. Results can be compared for different cups. For example, is the probability of a MacDonald's cup landing on its side greater than, less than, or equal to the probability of a Burger King cup landing on its side?<br><br>After the initial trials, permit the students to alter the cups by cutting off the top or bottom or by cutting holes in the side to get different results. |

TEACHER COMMENTS TO STUDENTS:

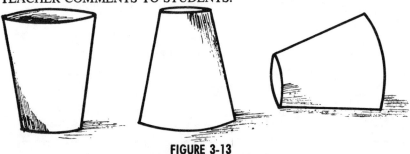

**FIGURE 3-13**

# Teaching Probability and Statistics

When a paper cup is tossed, there are three realistic ways it can land (outcomes): top, bottom, and side (Figure 3-13). Each time a toss is made is called a trial. After each trial the landing position is recorded by a tally mark for the particular position (top, bottom, or side in this example). After all trials are completed the tally marks for each position are counted and this total is called the frequency for that landing position. Every attempt should be made to throw the object in a similar manner for each trial. Complete the chart shown in Figure 3-14 for 30 trials of throwing a paper cup.

| Outcome | Tally | Frequency | Part of Total Trials |
|---|---|---|---|
| Top | | | |
| Bottom | | | |
| Side | | | |

**FIGURE 3-14**

Compare your results with those of others. Are the results similar? Tell why your comparison came out the way it did. Did the manner in which you tossed the cup influence the outcome?

Using your first trials as data, predict what would happen if you repeated the experiment.

- In___out of 30 tosses, the paper cup would land on top.
- In___out of 30 tosses, the paper cup would land on the bottom.
- In___out of 30 tosses, the paper cup would land on its side.

Now, using the same chart (Figure 3-15), test your predictions!

| Outcome | Tally | Frequency | Part of Total Trials |
|---|---|---|---|
| Top | | | |
| Bottom | | | |
| Side | | | |

**FIGURE 3-15**

**TITLE:**     **Baseball—Who's Worth It? (3.07)**
**GOAL:**     Students will become aware of the mathematical means by which the salary of a person may be determined on the basis of his productivity.

| | |
|---|---|
| OBJECTIVES: | The student will keep an accurate record of data collected. |
| | The student will form appropriate ratios with the data collected. |
| | The student will select the "ideal" indicator of productivity to determine the salaries of baseball players. |
| GRADES: | 6, 7, 8, General Mathematics. |
| MATERIALS: | Daily sports page. |
| INSTRUCTIONS FOR TEACHERS: | Motivate the activity by discussing high salaries paid to some players in the major leagues. |
| | Contact the news media, player's association, and/or major league offices for approximate salary figures for some players. |
| | Discuss the need for careful collection and recording of data. |
| COMMENTS: | This activity is ideal for baseball spring training since the duration of spring training falls entirely within the school year. |
| | You may wish to make a contest of this activity among members of the class. Remember, the highly paid stars may not be the most productive in terms of hits, stolen bases, or reaching base —especially during spring training. |
| | This may be altered to evaluate pitchers as well. |

TEACHER COMMENTS TO STUDENTS:

Choose a baseball player from any major league team and record his statistics throughout the spring training period. (See Figure 3-16.)

Player_____ Salary_____

| Date | Single | Double | Triple | HR | Fielder's Choice | Stolen Bases | Runs Scored | Errors |
|------|--------|--------|--------|----|--------------------|--------------|-------------|--------|
|      |        |        |        |    |                    |              |             |        |

**FIGURE 3-16**

*Teaching Probability and Statistics*

1. What's the ratio of your player's salary to the number of his base hits? Calculate the cost per hit by dividing the salary by the number of hits. This is rather realistic in the American League where another player bats for the pitcher.
2. What's the ratio of your player's salary to his scoring? If he gets on base but rarely scores, should his position in the batting order be changed?
3. What's the ratio of your player's salary to his fielding errors?
4. What other ratio would help your player support his salary demands?
5. How should players' salaries be determined?

**TITLE:** **Drop 'em (3.08)**

GOAL: We encourage students to conjecture, estimate and predict. Yet, seldom are the students provided with immediate data to check their endeavors. This activity will provide the opportunity for rapid verification of estimations.

OBJECTIVE: The student will predict the heights of a sequence of bounces of a given ball dropped from a known height.

GRADES: General Mathematics, Algebra II.

MATERIALS: Tennis ball, golf ball, handball, or basketball; smooth level surface next to a wall; paper and pencil.

INSTRUCTIONS FOR TEACHERS: This activity requires a smooth level surface next to a vertical wall. The students will drop a ball from a designated altitude and then, using successive horizontal sightings, estimate the height of each bounce. The number of bounces used can vary, but at least five should be used.

Different students should be assigned the task of mentally noting the height of the bounces and then, after the last recorded bounce, marking the height of their respective bounces. After a few trial drops the student will know about where his point will be. To eliminate distortion, the students should be encouraged to get on their knees, sit, or squat to provide a horizontal line of sight with the peak of each bounce. More than one recording trial should be made to check the accuracy of the results.

COMMENTS: One interesting related activity is to record bounce heights for different types of balls and then compare the results. Can bouncing of different balls be related? Is there a constant ratio between bounce heights one and two for the different balls? Does a constant ratio exist between bounce heights two and three for different balls?

For Algebra II students, you may wish to have them develop an equation which describes the curve generated by the points on the graph(s).

TEACHER COMMENTS TO STUDENTS:

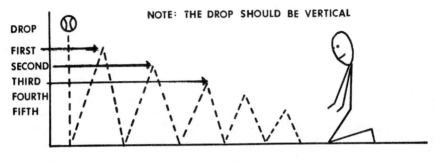

**FIGURE 3-17**

You are to drop a ball from a given altitude and record the height of the bounces (Figure 3-17). It will be to your advantage to drop the ball close to a vertical wall that is next to a smooth horizontal surface.

After the ball is dropped, different students should be responsible for recording different heights. A few practice drops will help determine the approximate vicinity you must watch when actually recording the results. Care should be taken to look horizontally when recording bounce heights. This horizontal sight line can be obtained by kneeling, squatting, or sitting, according to the height of the assigned bounce.

The heights of at least five successive bounces should be recorded. The ball should be dropped more than once and the results recorded to verify that the established heights are accurate.

After the results are finalized, graph the results, and draw a smooth curve through the points (Figure 3-18). Try it!

# Teaching Probability and Statistics

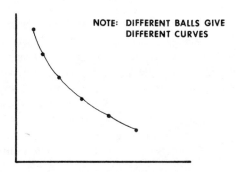

**FIGURE 3-18**

Compare the ratio between the height of the first and second bounce with the ratio of the height of the second and third bounce, etc. Is there a constant ratio that exists between the different bounces?

Using the information gathered from the first five bounce heights, try to predict the height of the sixth bounce.

If the ball were dropped from a point twice as high, would the successive bounces be twice as high? Try it!

Is the ratio between the successive bounce heights the same when the ball is dropped from different altitudes?

How are the graphs similar? Use this information to predict the curve for a third initial dropping altitude which is the average of the first two. Check your prediction by actually dropping the ball from this average height and graphing the resultant values. Try it!

**TITLE:** **Draw (3.09)**

GOAL: Rather than discussing probabilities and when they occur, students should have the opportunity to experience the concept development.

In this activity the student will draw objects out of a container, knowing the contents of the container. However, there will be a structured sequence to be followed as to what objects will be in the container.

OBJECTIVE: The student will, upon completion of this activity, be able to define probability.

| | |
|---|---|
| GRADES: | General Mathematics, Senior Mathematics. |
| MATERIALS: | Five objects of each of three colors but the same size and shape; opaque container that facilitates randomly removing objects from it; paper and pencil. |
| INSTRUCTIONS FOR TEACHERS: | Care should be taken that the students do the activity in the prescribed order, thus enhancing their ability to generalize. Furthermore, the effects of drawing objects with and without replacement should be clarified through discussion if necessary. If the students do not understand how to make generalizations, encourage them to repeat the activity, varying the number and colors. |
| COMMENTS: | Humor can be injected by having the elements be boys and girls, particularly when pairings are discussed. |

TEACHER COMMENTS TO STUDENTS:

You will be selecting from a container, objects which you cannot see. At times you will be instructed to return the objects to the container while at other times you will not return the object to the container after selection. In probability and statistics, these situations are referred to as selection with or without replacement, respectively.

Place five red objects in the container. If the favorable event is to draw a red object, the outcome of selecting a red object must always occur since only red objects are in the container. Check this by actually selecting objects from the container. In this situation the probability of selecting a red object is said to be *one*, while the probability of selecting a yellow object, for example, is said to be *zero*. This zero probability comes from an impossible situation, and in this case there are no yellow objects in the container so none could be selected.

Put one yellow object and one red object in the container. How many different colors can now be drawn from the container? This number is the number of possible outcomes. How many red objects could be drawn? The probability of a red is 1/2. Similarly, the probability of a yellow is 1/2.

This means that if you draw an object out of the container and then replace it, half of the draws would be red and the other half

*Teaching Probability and Statistics*

would be yellow. Make 20 draws with replacement to see how close to the expected probability you get (Figure 3-19). How close did you come to drawing half of the first ten draws as red?

| Draw | 1 | 2 | 3 | 4 | 5 | 6 | 7 | 8 | 9 | 10 | 11 | 12 | 13 | 14 | 15 | 16 | 17 | 18 | 19 | 20 |
|---|---|---|---|---|---|---|---|---|---|---|---|---|---|---|---|---|---|---|---|---|
| Color Drawn | | | | | | | | | | | | | | | | | | | | |

**FIGURE 3-19**

Now, put five red and five yellow objects in the container. Since half of the total number of objects is red, the probability of selecting a red object is 5/10 or 1/2. Again, the probability of selecting a yellow object would also be 1/2. Draw with replacement one object at a time from the container 30 times to see how closely you can get to a probability of 1/2 for red objects (Figure 3-20).

| Draw | 1 | 2 | 3 | 4 | 5 | 6 | 7 | 8 | 9 | 10 | 11 | 12 | 13 | 14 | 15 |
|---|---|---|---|---|---|---|---|---|---|---|---|---|---|---|---|
| Color Drawn | | | | | | | | | | | | | | | |

| Draw | 16 | 17 | 18 | 19 | 20 | 21 | 22 | 23 | 24 | 25 | 26 | 27 | 28 | 29 | 30 |
|---|---|---|---|---|---|---|---|---|---|---|---|---|---|---|---|
| Color Drawn | | | | | | | | | | | | | | | |

**FIGURE 3-20**

Using the terms "total possible objects" and "the number of objects of a given color," write a definition or description of probability. Convert this expression into symbols and you have a formula for calculating the probability of an event.

**TITLE:** **Color Wheel (3.10)**

GOAL: Many probability activities are oriented towards games and sports; thus, girls have little interest in them. This activity uses the painting of fingernails to gain the attention of girls for a probability exercise.

| | |
|---|---|
| OBJECTIVE: | The student will compute the probability of ten successive independent events. |
| MATERIALS: | Several different colors of fingernail polish, polish remover, tissues, newspaper, a spinner, large piece of cardboard. |
| GRADES: | 6, 7, 8, General Mathematics. |
| INSTRUCTIONS FOR TEACHERS: | This activity is best conducted in groups of three students. Boys may take part in this activity as recorders and painters. |
| | Be sure to put newspaper under the area where the painting will take place. Fingernail polish can be removed, but the remover may damage painted surfaces. |
| COMMENTS: | The Teacher Comments to Students section of this activity leads the student to the discussion of probability. However, there are many variations and extensions of this activity which will lead further into discussion of probability. |

1. What effect will different numbers of colors have on the success of guessing the color.
2. What is the probability that all nails will be red if red is one of the five distinct colors.
3. What is the probability that any pattern will occur if there are x distinct colors.
4. What happens to the probability if green is in two of the bottles in the set. That is, what happens if all x colors are not distinct?
5. Permutations and combinations could be discussed with algebra students.

TEACHER COMMENTS TO STUDENTS:

Who has the wildest colored fingernail polish in your class?

Let's try an experiment to see if we can make a set of fingernails even more wild than hers.

Get a spinner and a selection of fingernail polish colors. Be sure to let all members of your group see all the colors.

Now, select one person to conduct the experiment, one to record the results, and a victim.

Assign a number to each color.
1._____
2._____
3._____
4._____

Now, set the victim behind a shield so that she cannot see the spinner or the colors (Figure 3-21).

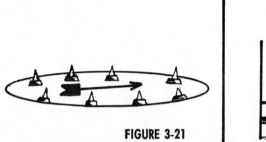

**FIGURE 3-21**

Let's begin the experiment. The victim must try to guess the color that the nail will be. Then spin the spinner and paint the nail the indicated color. Do all ten nails and record the results. (See Figure 3-22.)

| Right Hand | Little | Ring | Middle | Index | Thumb |
|---|---|---|---|---|---|
| Victim's Guess | | | | | |
| Color Painted | | | | | |

| Left Hand | Thumb | Index | Middle | Ring | Little |
|---|---|---|---|---|---|
| Victim's Guess | | | | | |
| Color Painted | | | | | |

**FIGURE 3-22**

How many guesses were correct?_____
How many guesses were there in all?_____
What is the ratio? correct guesses/guesses_____
How many do you think you could guess?_____
Try it!

# 4

# How to Make Everyday Numbers Rational

Mathematical concepts and skills related to the set of rational numbers are used daily by students in very meaningful ways. Experienced teachers are well aware of this fact, and they draw upon the students' experiences with whole numbers, integers, fractional numbers, decimals, ratio and percent to motivate the study of mathematics.

This chapter provides the teacher with examples of activities that will appeal to students of both sexes and of various interests. The rational numbers are ever present in the arts of sewing and cooking; thus, activities are provided which involve sewing in interior decorating, sewing clothing, clothes size, and cooking. The activity, No Chocolate Mess, is an example of a cooking activity which can be done in the mathematics classroom. Hopefully, the creative teacher will be motivated to develop many more activities involving sewing and cooking.

Of course, not all students are interested in domestic arts, but the concepts of rational numbers are also used by students in art projects, crafts, sports, and building construction. There is at least one activity from each of these areas of interest in this chapter.

Teachers of mathematics realize the need for maintaining and improving students' computational skills with the rational numbers and the major subsets of the rationals. To aid the teacher concerned with this goal, several games, tricks, and puzzles have been provided in this chapter. These reinforcement activities do not provide the teacher with an endless list of such activities, but the resourceful teacher will be able to use these activities as models to develop many more valuable games and puzzles.

**TITLE:** **By Nines in Ten (4.01)**

GOAL: Students are often intrigued by unique or "gimmicky" approaches to topics. Multiplication by nines in base ten on the bead computer is a quick attention getter.

OBJECTIVE: Using an apparatus made from wire and beads, the student will be able to multiply any single digit by 9.

MATERIALS: Coat hanger wire, five red beads, five blue beads, file, pliers.

GRADES: 4, 5, 6, 7, 8.

INSTRUCTIONS FOR TEACHERS: See Facto Beads, activity 4.02, for how to make the calculator.
Be sure to check the rules before the student gives them to another person to try.

COMMENTS: This pattern can be extended through 20 if 11 is omitted, through 30 if 21 and 22 are omitted, etc.
If multiplying 9 by 12 through 20, the leftmost bead is the hundreds, and the number of beads between that leftmost bead and the bead counted to by the ones digit of the "teen" number represents the tens digit of the product, while the number of beads to the right of the bead counted to gives the ones digit of the product, as shown in Figure 4-1 for 9 × 13. To get the answer, you count to the third bead. (9 × 13 = 117.)

hundreds      tens      ones

**FIGURE 4-1**

For 9 × 17, count to the seventh bead (Figure 4-2): 9 × 17 = 153.

How to Make Everyday Numbers Rational 83

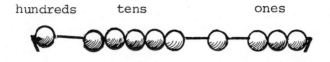

**FIGURE 4-2**

Skipping 21 and 22, factors of 23 through 30 multiplied by 9 can be done on the calculator in a similar manner.

For example: 9 × 25. The first two beads give the hundreds digit; the number of beads between the first two and the fifth bead gives the tens digit of the product. The number of beads to the right of the fifth bead gives the ones digit of the product.

For 9 × 25, count to the fifth bead (Figure 4-3): 9 × 25 = 225.

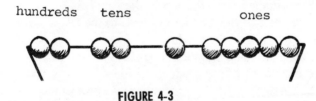

**FIGURE 4-3**

For 9 × 27 (Figure 4-4), count to 7.

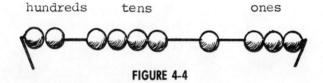

**FIGURE 4-4**

TEACHER COMMENTS TO STUDENTS:

The calculator used in Facto Beads is to be used here.

You will be multiplying single digits by 9. To do 9 × 4, place all the beads at the left, count over to the fourth bead and slide the rest (6 of them) to the right. Slide that fourth bead so there is space between it and the sets of beads at either end of the calculator as shown in Figure 4-5. The beads to the left of the fourth represent the tens digit of the product and the beads to the right of that fourth digit represent the ones digit of the product. So the product of 9 and 4 is 36.

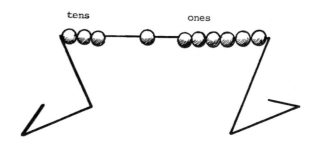

**FIGURE 4-5**

What did you do to multiply 4 by 9?
You counted from the left to the fourth bead.
You counted the number of beads to the left of the fourth bead for the tens digit of the product.
You counted the number of beads to the right of the fourth bead for the ones digit of the product.
To multiply 6 by 9 you would count from the left to the sixth bead and then to the left of it for the tens digit of the product and to the right of that sixth bead for the ones digit of the product.
Develop a series of rules that tell the steps that must be used when multiplying by 9 on this calculator. After you have checked to be certain those rules work, give your rules and the calculator to someone who has not seen or used the calculator and see if he can multiply 9 with it.

**TITLE:**     **Facto Beads (4.02)**

GOAL:     This activity, a variation of finger multiplication, is another example of the "gimmicky" approach that often appeals to students.

OBJECTIVE:     Using an apparatus made from wire and beads, the student will be able to find the products of one single digit 6 through 9 and a second single digit 6 through 9.

*How to Make Everyday Numbers Rational*

| | |
|---|---|
| MATERIALS: | Coat hanger wire, five red beads, five blue beads, file, pliers. |
| GRADES: | 4, 5, 6, 7, 8, |
| INSTRUCTIONS FOR TEACHERS: | The students should not do 7 × 8 as one of their first examples because they need to become familiar with the rules and the 7 × 8 situation will yield the correct response even when the rules are reversed. |
| | After the coat hanger wire is cut, be certain that the ends are filed to remove sharp spots. |
| | Be sure to check the rules before the student gives them to another person to try. |
| COMMENTS: | This can be either an individual or group activity and can be related to Activities 2.09 and 4.01. The rules used here are rather standard, but have been extended by Louisa R. Alger in "Finger Multiplication," *The Arithmetic Teacher*, April, 1968, pp. 341 - 43. |
| | Note that the bead color is insignificant. |

TEACHER COMMENTS TO STUDENTS:

Do multiplication facts give you problems? This activity will give you a way of getting the answer to multiplication facts which are usually hard to remember.

You will be given 10 beads, 5 each of 2 colors. You are to make a holder for these beads. The holder should look something like the one in Figure 4-6. The holder should be about 5 centimeters longer than 10 times the diameter of one bead. Put the 5 red beads on first and then 5 blue beads. Be sure all the beads are on before bending both ends of the frame.

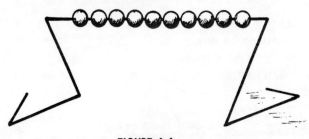

**FIGURE 4-6**

Suppose you wanted to multiply 9 × 6. Move 4, (9−5) red beads and 1, (6−5) blue bead to the center so all 5 touch and there are 3 distinct groups of beads, as shown in Figure 4-7.

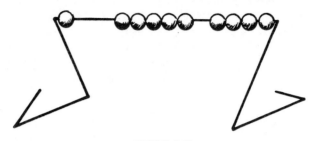

**FIGURE 4-7**

Multiply the 5 beads in the center by 10 and add that to the product of the 1 red and 4 blue beads. You now have 5(10) + 1(4) = 50 + 4 = 54.

Briefly, what did you do?

$$9 - 5 = 4 \qquad\qquad 6 - 5 = 1$$
$$4 + 1 = 5$$
$$5 \text{ times } 10$$
$$1 \text{ times } 4$$
$$50 + 4 = 54.$$

Do 7 × 7 with the calculator.

$$4 \text{ times } 10$$
$$3 \times 3$$
$$40 + 9 = 49$$

Develop a series of rules that tell the steps that must be used when multiplying on this calculator. After you have checked them to be sure those rules work, give your rules and the calculator to someone who has not seen or used the calculator and see if he can multiply.

**TITLE:** **Rise and Run (4.03)**

GOAL: Slope is often a difficult concept for students to understand. Part of this difficulty arises from the ap-

|                      |                                                                                                                                                                                                                                 |
|----------------------|-----------------------------------------------------------------------------------------------------------------------------------------------------------------------------------------------------------------------------------|
|                      | parent lack of practical use for slope. In this activity the concept of slope is explored in the world of carpentry.                                                                                                            |
| OBJECTIVES:          | The students will measure the rise and run of stairs. The students will compute the ratio of the rise to the run. The students will interpret the meaning of ratios.                                                            |
| MATERIALS:           | Rulers, available stairs.                                                                                                                                                                                                        |
| GRADES:              | 8, General Mathematics, Algebra.                                                                                                                                                                                                 |
| INSTRUCTIONS FOR TEACHERS: | This activity is best conducted as an independent study for small groups of students since large groups in the halls are often disturbing to others in the school. Also, in a large group not every student will participate actively in the project. |
| COMMENTS:            | For algebra students and other bright students the factor of the inclination of stairs with constant ratios of rise to run may occur. For your information, the ideal inclination of stairs is between 30 degrees and 35 degrees. |

TEACHER COMMENTS TO STUDENTS:

Have you ever noticed that climbing some stairs is more difficult than others? Well, why is this so? List your reasons below.
1. _____
2. _____
3. _____
4. _____
5. _____

Let's look more closely at the construction of stairs! Go to the stairs that your teacher has selected, and measure as shown in Figure 4-8.

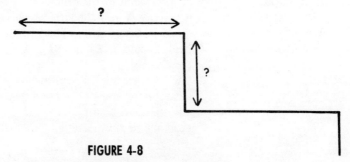

**FIGURE 4-8**

The vertical distance is called the rise of the stair and the horizontal distance is the run of the stair.

Look at other stairs in the school. Measure the rise and run of these stairs. Do all stairs seem to have the same rise and run?

Think about the rise and run for a moment! What would happen to the stairs if the rise were increased and the run held constant?

What would happen if the rise were held constant and the run were increased?

Consider the following ratios of rise to run:

$$\frac{\text{rise}}{\text{run}} = \frac{12 \text{ in.}}{12 \text{ in.}} = \frac{24 \text{ in.}}{24 \text{ in.}} = \frac{36 \text{ in.}}{36 \text{ in.}}$$

Which would describe the best stairs?_____Why?_____

_____

What ratio of rise to run do you think makes the best stairs?_____Why do you think so?_____

_____

| | |
|---|---|
| **TITLE:** | **Balloon Animals (4.04)** |
| GOAL: | Students enjoy working with materials which are not usually associated with mathematics. This activity uses balloons to help children communicate this knowledge of fractions. |
| OBJECTIVES: | The students will compute the average of a given set of numbers. |
| | The students will write directions for making a balloon animal, using fractions in their descriptions. |
| MATERIALS: | Balloons (the long ones). |
| GRADES: | 6, 7, 8, General Mathematics. |
| INSTRUCTIONS FOR TEACHERS: | This activity is best conducted with small groups of students. The number of balloons given to the students should be controlled until the last phase of the activity. |
| | As the "wiener dog" is constructed, use fractions to describe the steps of construction. For example, "Betty used about 1/3 of the long balloon for the head and neck of the dog." If the class has a poor |

background in fractions, you may wish to demonstrate 1/2's, 1/3's, 1/4's, etc., to the class, until the parts become very apparent.

TEACHER COMMENTS TO STUDENTS:

**FIGURE 4-9**

Take one balloon, blow it up using big breaths, and keep a count of how many breaths are needed to fill the balloon.

Write your count of the number of breaths needed to fill the balloon in the chart on the chalkboard. Who used the most?_____ Who used the least?_____ What is the average for the class? _____

Let a little air from your balloon and tie the ends. Now let's make a "weiner dog" (Figure 4-10).

**FIGURE 4-10**

How many balloons did you use?_____
Who in the class used the most balloons?_____
How many?_____Who used the least?_____
How many?_____ Now, get some balloons and make another animal of your choice.

Write up the directions for making your animal. Use fractions in your directions!

### TITLE: 100% Mobile (4.05)

| | |
|---|---|
| GOAL: | This activity uses a simple artistic device to show equivalent decimals, fractions, and percents. |
| OBJECTIVES: | The students will write equivalent percents and decimals for given fractions. |
| | The students will make a mobile to show equivalent values of percent, decimals, and fractions. |
| MATERIALS: | Three pieces of different colored construction paper per student, coat hangers, string. |
| GRADES: | 6, 7, 8, General Mathematics. |
| INSTRUCTIONS FOR TEACHERS: | This activity is ideal as a culminating lesson on the topic of equivalent values of percents, decimals and fractions. The students may work better in pairs for this activity due to the extra hands needed to make a mobile. The completed mobiles will not only serve an instructional purpose, but can also be very decorative in the classroom. |

TEACHER COMMENTS TO STUDENTS:

Get three pieces of 8½ × 11 construction paper which are three different colors, and cut and label each of them as shown in Figure 4-11.

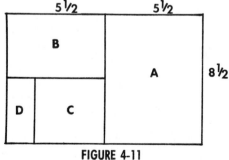

**FIGURE 4-11**

*How to Make Everyday Numbers Rational*

Choose a color.

A is what fractional part of the rectangle?_____
  Write this fraction on A.
B is what fractional part of the rectangle?_____
  Write this fraction on B just as you wrote 1/2 on A.
C is what fractional part of the rectangle?_____
  Write this fraction on C.
D is what fractional part of the rectangle?_____
  Write this fraction on D.

Compare the three pieces labeled A. Do they match up exactly? That is, are they congruent?_____

Write the decimal for 1/2 on another A.

Write the percent for 1/2 on the last A.

Using the same color for decimals, write the decimal values of 1/4 on B, 3/16 on C, and 1/16 on D.

Using the same color for percents as you did for 50%, write the correct percent values on B, C, and D.

Now make a mobile with your pieces of construction paper. Start with the big pieces! 1/2 = 50%. (See Figure 4-12.) Do they balance? What does this tell you?_____

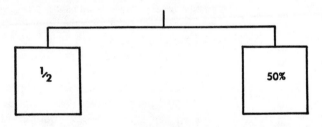

**FIGURE 4-12**

Add two more pieces to the mobile so that the colors are the same as the first two pieces. Do they balance? _____

Put all four of one color on one side and all four of another color on the other side. Do they balance? _____

What is the sum of the value of the pieces on the left?_____

What is the sum of the value of the pieces on the right?_____

Since these two sides are balanced, what do you know about the sums?_____ Thus, _____ = 100%. What would be the same as 100% if you used the other color? _____

**TITLE:** **Top Speed of a Ten-Speed (4.06)**

**GOAL:** The advantages of ten-speed bicycles can be seen when going up hills. However, students frequently wonder if the top speed of a ten-speed is any "faster" than that of other bicycles.

**OBJECTIVE:** The student will calculate the speed, in miles per hour, of the different speeds of a ten-speed bicycle.

**GRADES:** 7 - 12.

**MATERIALS:** Ten-speed bicycles, stopwatch, linear measuring tape.

**INSTRUCTIONS FOR TEACHERS:** The student will turn the bicycle upside down and pedal at a constant rate in tenth gear, noting the number of revolutions the back wheel makes while being driven. Since they can measure or compute the circumference of the back wheel, the students can then calculate the distance the bicycle could go in a time unit by multiplying the circumference by the number of revolutions. Thus, the speed could then be converted to miles per hour. This could be repeated for each speed on the bicycle.

The students should then ride the bicycle through an established level distance at the same pedaling rate and compare this speed with the "upside-down" speed of the tenth gear. Questions such as: "Is this ratio between true speed and upside-down speed the same for all gears?" can be developed.

**COMMENTS:** Perhaps the students could compare top speeds for different brand ten-speed bicycles or bicycles that are not all ten speeds. They could count the number of teeth on each of the sprockets and establish ratios as to the number of teeth on the two gears involved. Is there a constant relationship between the second and third speeds, third and fourth, etc.?

TEACHER COMMENTS TO STUDENTS:

The objective in this experiment is to calculate the "theoretical" top speed of a ten-speed bicycle and compare it with the actual speed of the bicycle when ridden on smooth level ground at the same pedaling rate used for the theoretical top speed. The first part of this experiment requires that the bicycle be turned upside down. Suppose a realistic pedaling rate is 20 complete revolutions of the pedals per minute on smooth level ground. At that rate, how far would the bicycle travel in one minute?

This value can be derived by measuring the number of revolutions the rear wheel makes in one minute at the pedaling rate and then multiplying that number times the circumference of the wheel:

>Number of revolutions of rear wheel _____.
>Circumference of rear wheel _____.
>Distance theoretically traveled by rear wheel at given rate _____.

This could then be converted to miles per hour for the speed as it is usually listed. Your converted speed in mph is _____.

Take the bicycle to a smooth level surface and, pedaling at the same rate, have a rider travel a specified distance. This distance and the time required to traverse it can be converted to speed for this pedaling rate.

1. Trial one actual top speed _____.
2. Trial two actual top speed _____.
3. Trial three actual top speed _____.
4. Trial four actual top speed _____.
5. Trial five actual top speed _____.
6. Average actual top speed _____.
7. Theoretical top speed _____.

If the two speeds are not the same, what would cause the difference? _____

Is it possible to bias the outcome? _____

If you repeated this procedure for 9th gear as well as 10th, is the difference between 9th and 10th gear "real" speed? Why or why not?

## TITLE:      Pieces of % (4.07)

GOAL:      This activity allows students to determine "parts of one hundred" on the concrete level.

| | |
|---|---|
| OBJECTIVES: | The student will manipulate pieces of graph paper to indicate "parts of one hundred." |
| | The student will write and state parts of one hundred as both decimals and percent. |
| GRADES: | 6, 7, 8, General Mathematics. |
| MATERIALS: | One ten by ten centimeter of graph paper, several smaller pieces of graph paper cut to represent various percentages. |
| INSTRUCTIONS FOR TEACHERS: | Students should be encouraged to answer each question before going on. |
| COMMENTS: | If millimeter graph paper is available, the standard piece of graph paper can be made ten centimeters by ten centimeters with the millimeter marks as subdivisions. This will allow the inclusion of percents such as .4%, 20.5%, etc. |
| | Percent is related to fractions in this activity, but it could be rewritten easily to correlate decimals and percents. |

TEACHER COMMENTS TO STUDENTS:

Get three pieces of graph paper. Now, cut out these smaller pieces of graph paper and number them as you cut them.

1. 10 centimeters by 10 centimeters square.
2. 1 centimeter square.
3. 10 centimeters by 10 centimeters square with one 1 cm. square removed.
4. rectangle which is 1 centimeter by 4 centimeters.
5. 5 centimeters by 5 centimeters square, and
6. rectangle which is 3 centimeters by 10 centimeters.

Use piece No. 1 as your standard shape. Place the smaller pieces of graph paper on top of this paper one by one and fill in the chart as shown in Figure 4-13.

| Piece | Number of Squares | No. of Squares in Piece #1 | Fraction of #1 Covered |
|-------|-------------------|----------------------------|------------------------|
| 2 | | | |
| 3 | | | |
| 4 | | | |
| 5 | | | |
| 6 | | | |

**FIGURE 4-13**

There is another way to write fractions with a denominator of 100. The sign, %, means part of 100 and is read "percent." So, piece 2 represents 1/100 or 1% of piece 1.

Write the fractions for pieces 3, 4, 5, 6, as percents.

Piece 3     99/100   = _____   Piece 4 _____ = _____
Piece 5 _____     = _____   Piece 6 _____ = _____

Now, using a clean sheet of graph paper, make pieces of paper that show the following:

11%, 23%, 85%, 7%, 45%, 36%, and 66%.

### TITLE: "Lights; Camera; Action!" (4.08)*

GOAL: Students often have difficulty understanding and/or remembering the sign of the product when multiplying two signed numbers. This activity will aid in this necessary memory task while also providing a vehicle useful in investigating division of two signed numbers.

OBJECTIVE: The student will recall the rules of multiplying integers.

GRADES: 6 - 12.

MATERIALS: Movie camera (8mm cost is much less), movie projec-

---

* Initial idea by Merle Grady, University of Dallas.

tor (must be reversible - stop action not necessary), one roll of unexposed film, students.

INSTRUCTIONS FOR TEACHERS:
The students will be filmed while walking. Initially, they are to indicate the direction of their walk (hand in front for forward; hand in back for backward). The students can be filmed as they walk in a row, column, circle, etc. The last part (at least a third) of the film should not have the direction of walk indicated.

Upon development, the forward walk depicts a positive value while the backward walk depicts a negative value. While showing the film, the projector in forward represents positive while reverse is negative. The direction of the walk and the motion of the projector are the two factors signs and the sign of the product is determined by the direction of the walk on the screen.

The projector in forward (positive) showing students who were walking forward when filmed (positive) will give a picture on the screen of the students walking forward (positive). So, this depicts that a positive times a positive gives a positive result: $(+) \times (+) = (+)$.

The projector in forward (positive) showing students who were walking backward when filmed (negative) will give a picture of the students walking backward (negative). This is shown as: $(+) \times (-) = (-)$.

The projector in reverse (negative) showing students who were walking forward when filmed (positive) will give a picture of the students walking backward (negative): $(-) \times (+) = (-)$.

The projector in reverse (negative) showing students who were walking backward when filmed (negative) will give a picture of the students walking forward (positive): $(-) \times (-) = (+)$.

The part of the film where the students do not indicate the direction of their walk can be used to introduce division of signed numbers. The students should be familiar with projector directions (one factor) and walk direction as shown on the wall (net result or product) from the earlier activity. Knowing or determining the direction of the projector and determining the resultant direction of the walk as shown as the wall, the student should be able to determine the direction of walk when initially filmed: a representation of the missing factor.

COMMENTS: This activity provides a sometimes necessary mnemonic device for remembering something. The cost of film, availability of equipment and speed of development aid in making this a worthwhile activity. Most people enjoy seeing movies of themselves or their friends and a group of students almost always want to watch it again for the entertainment factor. However, while they are being entertained, the point is being re-emphasized.

Conceivably, one student could direct the filming operation, another film the action, a third run the projector, a fourth be responsible for getting the film developed, etc., each of which increases student involvement.

TEACHER COMMENTS TO STUDENTS:

You will participate in the making of a movie which will help you remember the sign of the product of two signed numbers and the sign of the quotient of two signed numbers. For the movie, define a forward walk as positive and a backward walk as negative. The projector in forward represents positive, while reverse is negative. These situations (walking direction when filmed and projector direction) represent the two factors. The image as projected on the screen will show either a forward walk (positive) or backward walk (negative), representing the product.

For at least the first part of the movie, those being filmed should indicate the direction they are walking—first forward (10 to 12 feet of film) and then backward for approximately the same amount of film—perhaps by holding their hands in front or behind themselves while walking or by having some students wear baseball caps with the peak in front or in back. The last part of the film should be done with the students not indicating the direction of their walk. In this part of the film, spend shorter periods of time filming each direction and do both directions more than once.

After the film has been developed, it will be a helpful aid for determining and remembering the sign of the product of two signed numbers. When the projector is running forward showing the first segment of film where the students were filmed walking forward, the figures on the screen will be walking forward. This can be expressed symbolically as a positive (walking forward) times a positive (projector running forward), giving a positive (images on the screen walking forward); that is, $(+) \times (+) = (+)$. This could also be expressed in chart form in at least two ways (Figure 4-14).

|  | Walking Direction | |  | + | − |
|---|---|---|---|---|---|
| Projector Direction | forward | backward |  |  |  |
| forward | forward |  |  | + | + |
| backward |  |  |  | − |  |

**FIGURE 4-14**

Look at the remaining part of the film in which the direction of walk is indicated, and complete both charts before looking at the part of the film not having the direction of walks shown.

When looking at the rest of the film, you will not know what direction the people were walking when they were filmed. In other words, the sign of one of the factors will be unknown. However, the walking direction of the images on the screen will indicate the sign of the product and the direction of the projector will give the sign of one of the factors. Since the other factor is missing, the situation is representative of the division of signed numbers.

Suppose the projector is running forward and the images on the screen are walking backward. What should the direction of the walk have been when the individuals were filmed initially? They had to have been walking backward because the only way the product can be negative is to have one of the factors be negative and recall that positive symbolizes forward motion and negative represents backward. In chart form:

*How to Make Everyday Numbers Rational* 99

| Direction of image on screen | Direction of projector | Walking direction when filmed |
|---|---|---|
| backward | forward | backward |
| backward | backward | |
| forward | forward | |
| forward | backward | |

Or;

| Product | Known factor | Missing factor |
|---|---|---|
| − | + | + |
| − | − | |
| + | − | |
| + | + | |

Now, having watched the rest of the film, complete both of the above charts.

**TITLE:** **Equations (4.09)**

GOAL: Students are often motivated to practice computation when the exercises are presented in puzzle form. This puzzle is designed for practice in the four operations on whole numbers and the order of operations.

OBJECTIVE: The student will identify equations involving whole numbers presented in a puzzle form.

GRADES: 7, 8, 9.

MATERIALS: Puzzle sheet.

INSTRUCTIONS FOR TEACHERS: Duplicate the puzzle sheet and directions for the students.

COMMENTS: Many of these puzzles may be made rather quickly; this puzzle is merely provided as an example. For brighter students, similar puzzles can be designed for fractional numbers, integers, and rational numbers expressed as decimals.

TEACHER COMMENTS TO STUDENTS:

Equations are present in this puzzle in vertical, horizontal and diagonal form. Look at the puzzle closely! How many different equations can you find?

Insert the signs in the equations and circle them as shown in Figure 4-15.

| 23 | 76 | 25 | 56 | 28 | 6  | 20 | − 8 | = 12 |
|----|----|----|----|----|----|----|-----|------|
| 4  | 28 | 1  | 28 | 28 | 5  | 4  | 7   | 5    |
| 53 | 3  | 23 | 28 | 9  | 2  | 3  | 1   | 6    |
| 2  | 8  | 7  | 5  | 3  | 6  | 5  | 6   | 5    |
| 9  | 2  | 4  | 41 | 7  | 24 | 36 | 3   | 2    |
| 7  | 36 | 3  | 4  | 2  | 30 | 1  | 6   | 7    |
| 63 | 6  | 5  | 4  | 4  | 3  | 21 | 4   | 14   |
| 5  | 6  | 42 | 32 | 23 | 9  | 14 | 8   | 6    |
| 9  | 6  | 3  | 81 | 8  | 3  | 3  | 42  | 5    |

**FIGURE 4-15**

**TITLE:** No Chocolate Mess (4.10)

GOAL: In class fun activities can be good motivators. Once the students are familiar with the results of this activity there will undoubtedly be many requests for repeats.

OBJECTIVE: After successfully completing a teacher designated review, small groups of students will make chocolate-covered peanut butter balls.

MATERIALS: Peanut butter, powdered sugar, butter or margarine, milk chocolate, paraffin, stove or hotplate or fondue pot, cookie sheets and bowls.

GRADES: 6 - 12.

INSTRUCTIONS FOR TEACHERS: The recipe for the peanut butter balls is:

    1 unit peanut butter
    1 unit powdered sugar
    ¼ unit margarine or butter
    ½ unit chocolate
    ⅛ unit paraffin

Step 1—Put the peanut butter in a bowl and gradually hand mix in the powdered sugar and melted margarine or butter. Once it is mixed thoroughly, shape the peanut butter dough into small balls. The diameter of these can vary depending upon the desired ratio of the amounts of chocolate to the amount of peanut butter mix. A ½" diameter is suggested.

Step 2—Melt the chocolate in a fondue pot or heavy pan at a low temperature, adding enough paraffin to make the chocolate smooth and "soupy."

Step 3—Dip the balls in the chocolate, then place them on a cookie sheet or table covered with waxed paper or aluminum foil to cool.

COMMENTS: A fondue pot could be used to heat the chocolate, and the balls dipped much the same as when fonduing. If a fondue pot is not used, it will be necessary to reheat the chocolate mixture when it cools.

There are no teacher comments to students for this activity.

**TITLE:** **How Much Material? (4.11)**

GOAL: Students need to be aware of which material will result in the smallest amount of scraps. In this activity, the student compares different fabric widths for the best buy.

OBJECTIVE: The student will compare the pattern yardage requirements for different material widths to determine which width is the best when desiring to keep scraps to a minimum.

MATERIALS: Pattern book or pattern envelope, tape measure.

GRADES: 6 - 12.

INSTRUCTIONS FOR TEACHERS: Ordinarily, fabric is folded with the selvage edges together before cutting out a pattern. Consideration should be given to your dimensions when using this procedure so that half the width of the fabric is not exceeded by the width of a pattern piece.

Although most patterns can be laid out on unfolded fabric, usually this process accomplishes little.

COMMENTS: It is assumed that the students will not be making a garment in this activity, but prevailing interest could result in finished products in a style show. Variations can be added through things like dresses with one side.

TEACHER COMMENTS TO STUDENTS:

If you do not know your chest measurement, have someone take it for you. Looking either in a pattern book or on a pattern package, determine your size. If your measurement is between sizes, buy the smaller size. Note that most patterns will need to be adjusted somewhat if the pattern is to fit your body well.

Once you have determined your pattern size, select a pattern. Look in the catalog for a pattern you like and look at the required fabric on the back of the pattern envelope section for that pattern. A sample is listed in Figure 4-16.

| Vest and pants: | | | | | | | | |
|---|---|---|---|---|---|---|---|---|
| Bust | 31 | 32 | 33 | 34 | 30½ | 31½ | 32½ | 34 |
| Sizes | 5jp | 7jp | 9jp | 11jp | 6 | 8 | 10 | 12 |
| 35" w/wo nap | 2⅝ | 2¾ | 2¾ | 2⅞ | 2⅞ | 3 | 3 | 3 |
| 45" w/wo nap | 2⅛ | 2⅛ | 2⅛ | 2¼ | 2⅜ | 2⅜ | 2⅜ | 2⅜ |
| 60" w/wo nap | 1⅝ | 1⅝ | 1⅝ | 1⅝ | 1⅝ | 1¾ | 1¾ | 1¾ |

**FIGURE 4-16**

Notice that for size 5 junior petite (5jp) the fabric needed for 45" is ½ yard less than the fabric needed for 35" material while the width difference is 10 inches. Similarly, ½ yard less is needed in 5 jp when using 60" material rather than 45", but the difference in the width is 15 inches. Why would the greater width difference not result in a smaller amount of needed fabric? In the 5jp, the difference in fabric between listed widths is ½ yard. Is this true for all the sizes?_____ If no, list the size and the differences._____

How much does the required yardage change from fabric width to fabric width for your size?_____

Notice that 5jp requires 2⅝ yards of 35" material and size 6, which has a bust size ½ inch smaller, requires ¼ yard more 35" material. The 45" material shares the same ¼ yard difference in favor of the size 6 while the 60" material shows no difference in the amount of required fabric for the two sizes. Compare size 7jp with size 8, 9jp with 10, and 11jp and 12 for differences in amount of fabric needed._____

Fabric is usually sold by the yard length. Advertisements will list the width of the fabric. Check at a fabric store or in newspaper ads to see if the wider material is more expensive. For example, does the 60" material cost 1⅓ times that of a yard of 45" material? Considering that there is less material required to make a garment as the width of the fabric increases, would it be smart to look for something in the wider fabric?

Using your size and comparable fabric of your choice that is available in the listed three widths, compute the cost of making the pants and vest listed earlier.

| Size | Fabric Needed | Cost per Yard | Total Cost |
|---|---|---|---|
| 35" w/wo nap | | | |
| 45" w/wo nap | | | |
| 60" w/wo nap | | | |

Which is the best buy?_____

# 5

# Mathematics Activities for Daily Living

Mathematics occurs in virtually every career or aspect of daily life, as you are undoubtedly well aware. However, students frequently fail to grasp the significance of mathematics in their lives and often question the necessity of learning particular mathematical concepts or for developing mathematical skills. These activities are designed so that the teacher of mathematics can show students places where mathematics is applied. The diversity of the activities in this chapter will assist the resourceful teacher in devising activities which would appeal to many students.

Career education, much broader than vocational, should be the concern of all teachers of mathematics since students should be prepared to cope with daily situations as a part of being a citizen, person, consumer, member of a household, owner of property, etc. To be effective in any of these roles, individuals must be capable of employing wise decision-making processes, spending procedures, effective use of leisure time, etc. These activities are designed to provide the student with background information which forms the basis of his/her fundamental life style. The students will do comparative shopping, practice decision-making, experience dealing with money, and engage in activities relating to science and mathematics. The concerned teacher of mathematics will want to use these activities as a source of ideas for developing many career and consumer activities for students.

**TITLE:** **The Best Price (5.01)**

**GOAL:** With today's versatile shopping facilities, the wise consumer avails herself of the best "bargains" for her money. This activity provides the student with a means of determining the store that offers the best buy for the money.

**OBJECTIVES:** The student will select a list of grocery items commonly used by his/her family and compare the prices of those items in different stores.

**MATERIALS:** Clipboard, shelf paper, or plain newsprint.

**GRADES:** 8, General Mathematics.

**INSTRUCTIONS FOR TEACHERS:** You need to help the students determine the stores to visit and the items to use on their list. They need to review the difference between "name" brand and "store" brand. One final reminder that should be given to the students is to compare the same size container between all stores.

**COMMENTS:** This activity is not practical for more than small groups. Careful briefing would permit sending different members to different stores with a prepared list of items.

TEACHER COMMENTS TO STUDENTS:

You are to prepare a list of grocery items your family normally uses and find which grocery store you should use to buy that list of items at the best price. The items and stores should be placed in a chart similar to the one shown in Figure 5-1. List the price for each item from each store in the right place. Total each column under each store and see which store gives the lowest total for all the "name" brand items and for the "store" brand items. Use the same brand for an item in each store.

|  | Store A | | Store B | | Store C | |
|---|---|---|---|---|---|---|
|  | brand | store | brand | store | brand | store |
| Bread |  |  |  |  |  |  |
| Milk |  |  |  |  |  |  |
| Potatoes |  |  |  |  |  |  |

**FIGURE 5-1**

## TITLE: Buy Right (5.02)

**GOAL:** Students need to become more aware of the uses of mathematics in the world of the consumer. Although stores now list comparative prices, many people do not use them, are not aware of them, or do not know how they are derived.

**OBJECTIVE:** The student will use proportions to compare the retail prices of different size containers of a product.

**MATERIALS:** Different size containers of a product and the price of each.

**GRADES:** 7, 8, General Mathematics.

**INSTRUCTIONS FOR TEACHERS:** The student needs to realize that he is getting a price for one unit (usually ounce) for each size and then compare those prices to see which is the best value.

**COMMENTS:** When the student considers the best value he needs to look at related influencing factors. For example, perhaps the larger size is less expensive per unit but, if it is not all used before spoiling, is it practical to purchase the larger size?

Although this can be done either as an individual activity or as a group the motivation for this activity could be enhanced by visiting a store. Newspaper advertisements could be used if a store trip is not practical. Newspapers could also be used to make comparisons between different stores that list the same brand but different sizes.

### TEACHER COMMENTS TO STUDENTS:

Are you a wise shopper? Is the largest size of an article always the best buy?

Select an item which comes in at least three sizes and note the price of each.

To find the cost per ounce, divide the number of ounces of product (net weight) into the cost.

The cost per ounce of the small size is _____.
The cost per ounce of the medium size is _____.
The cost per ounce of the large size is _____.
Which product is the best buy?

| | |
|---|---|
| **TITLE:** | **The Street Value of Tobacco (5.03)**[3]* |
| GOAL: | The students will determine the street or retail value of tobacco. |
| OBJECTIVES: | The student will weigh the tobacco in one cigarette. The student will compute the amount of tobacco in a pack of cigarettes. The student will compute the cost of one ounce of tobacco. |
| GRADES: | 7 - 12. |
| MATERIALS: | One pack of cigarettes for each group and balance scales. |
| INSTRUCTIONS FOR TEACHERS: | This laboratory activity is designed to take advantage of the present concern with drug abuse. The street value of drugs is often an integral part of news stories concerning the confiscation of illegal drugs by law enforcement officials. However, few students, or adults for that matter, have thought about the meaning of the term "street value." Essentially, this term refers to the retail value of the drug, but most mathematics teachers will agree that "retail sales" is not the most motivating mathematical topic. You will find that a discussion of the street value of tobacco, compared to the reported street value of many illegal drugs, will motivate students. |
| COMMENTS: | This activity is basically designed for one day, but it may be expanded easily. Library research can be assigned to determine the wholesale value of tobacco or the retail value of tobacco sold in the United States in one year. Determination of comparative costs of tobacco in regular cigarettes, filter cigarettes, cigars and pipe tobacco is another interesting activity. Perhaps law enforcement officials could be invited to bring in some confiscated drugs to be weighed and priced. This activity can serve the additional purpose of exposing the students to the law enforcement officers in a positive manner. |

---

[3]Brumbaugh, D.K., and Hynes, M.C., "Math Lab Activities: The Street Value of Tobacco," *School Science and Mathematics*, January 1974, pp. 75-76.

* Initial idea posed by William E. Brant.

## TEACHER COMMENTS TO STUDENTS:

How much tobacco is contained in one cigarette? Carefully remove the tobacco from one cigarette (standard size, non-filter) and weigh it on the balance scales.

One cigarette contains_____oz. of tobacco.

The tobacco in one pack of cigarettes weighs_____oz. DO NOT weigh any more tobacco to find your answer!

How many cigarettes are needed to obtain one ounce of tobacco?_____

Knowing there are 20 cigarettes in a pack, and using the results from the number of cigarettes in an ounce, determine how much tobacco is in a pack of cigarettes.

Is there a difference in your two computations of the amount of tobacco in a pack of cigarettes? If so, why do you think this difference appeared?

Now take the paper off all the cigarettes in your pack and carefully weigh the tobacco. How does this answer compare to your two computed answers?

Using the weight of the tobacco in a pack of cigarettes, determine the following:

1. What is the cost per ounce of tobacco in a pack of cigarettes?
2. What is the cost per ounce of tobacco in a carton of cigarettes?
3. How does this price compare to the reported price for an ounce of marijuana?
4. How does this price compare to the reported price for an ounce of hashish?
5. Can you afford to smoke?

## TITLE: Check Monopoly (5.04)

GOAL: This activity is designed to allow students to practice business record keeping in a recreational situation.

OBJECTIVES: The student will write checks and deposit slips to record all transactions in the game of monopoly.

The student will add and subtract integers in dollar form.

The student will check all computations with a calculator.

| | |
|---|---|
| GRADES: | 5, 6, 7, 8, General Mathematics. |
| MATERIALS: | Monopoly game,[4] one checkbook per person, one book of deposit slips per person. |
| INSTRUCTIONS FOR TEACHERS: | Play the monopoly by the stated rules except:<br>1. all purchases are paid by check;<br>2. all receipts must be recorded as deposits in the checkbook;<br>3. a player may accumulate debts up to $5000 before he is bankrupt; and<br>4. each player must keep a log of the property owned and the income from each. |
| COMMENTS: | Since Monopoly is a long game, the laboratory activity may last for several weeks. The logs of property owned and the checkbook will facilitate the continuity of the game regardless of the temporary stopping periods.<br><br>If time is a factor, the teacher may set a time period to terminate the game, rather than bankruptcy. The log of rents from properties may serve another purpose. It is assumed that "Board Walk" is a desirable property to own; however, will the rent receipts support this assumption? Let the students share their experiences of owning different properties. On which properties would houses and apartments be most beneficial?<br><br>This activity has no Teacher Comment to Student section since the directions for Monopoly are available with the game. |

| | |
|---|---|
| **TITLE:** | **Your Change, Mister! (5.05)** |
| GOAL: | The frequent complaint of small businessmen who employ students is that the products of new mathematics cannot make change in business transactions. This activity is designed to overcome this difficulty. |
| OBJECTIVES: | The students will design a flow chart for the process of making change with the aid of the teacher. (See Figure 5-2 on page 112.) |

---

[4] Monopoly, Parker Brothers; Salem, Massachusetts, 1941.

|  |  |
|---|---|
|  | The students will make change in a laboratory setting with accuracy for amounts less than or equal to one dollar. |
| MATERIALS: | Real coins or play money. |
| GRADES: | 5, 6, 7, 8, General Mathematics. |
| INSTRUCTIONS FOR TEACHERS: | Many young people attempt to make change by subtracting the purchase price from the amount given by the customer. The difference is then extracted from the coins in the cash drawer. However, this procedure causes the student difficulty in counting back the change; and as a courtesy, the change should be counted back to the customer, not dropped in his hand. The existence of this deficiency in the mathematical skills of your students may be established by allowing the students to role-play the clerk-customer relationship. |
|  | The development of a flow chart for making change should be conducted by the student with the aid of the teacher. The chalkboard and overhead projector are effective visual aids for this procedure. |
|  | Once a flow chart is designed, the students should practice the procedure until the skill of making change is well developed. This practice can take place in an interest center designed around a store theme, can be demonstrated with real money using pairs of students, or can be accomplished by permitting the entire class to use play money and ditto sheets of transactions. |
| COMMENTS: | There is no Teacher Comments to Students section for this activity. |
|  | The flow chart for making change can be expanded to include situations where the purchase price may require that the customer give the clerk a five- or ten-dollar bill. |

**TITLE:** **What's the Difference? (5.06)**

| | |
|---|---|
| GOAL: | As the metric system is introduced, demands for familiarity with the system will be increased. |
| OBJECTIVE: | The student will compare mileage per gallon in the U.S. common system with the "mileage per gallon" in the metric system. |

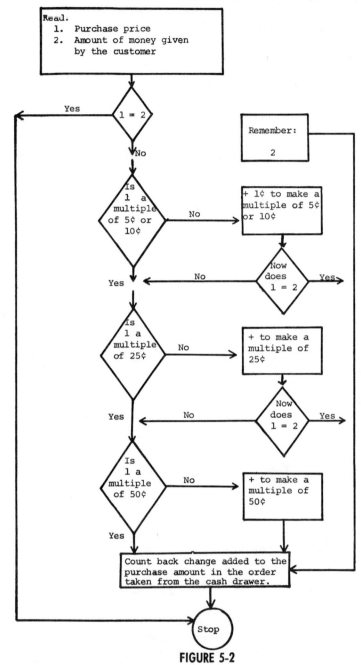

**FIGURE 5-2**

## Mathematics Activities for Daily Living

GRADES: General Mathematics.
COMMENTS: The students could make conversion charts listing comparative economy values for a car.

TEACHER COMMENTS TO STUDENTS:

Suppose two people are discussing the economy delivered by their cars. Sam's car delivers 20 miles per gallon and Pete's gives 32 kilometers per gallon. Whose car is the more economical, assuming there are 1.6 kilometers per mile?

**TITLE:** **Mobile (5.07)**

GOAL: There are several recreational uses of mathematics in our world of which students should be aware. This activity will show some of these uses and also involve some simple physics.

OBJECTIVE: Each student will build a mobile and balance corresponding pieces.

GRADES: 7 - 12.

MATERIALS: String, pipe cleaners, wire or coat hanger, pliers, picture supply (magazine, newspaper, photos, etc.).

INSTRUCTIONS FOR TEACHERS: The discussion about this mobile can be centered around geometric shapes as they appear in our everyday world. Students will be asked to choose geometric pictures from provided magazines. The picture will be placed on the mobile and balanced with a model of the picture. The student's concept of various geometric shapes will be utilized in making pipe cleaner models of the pictures. There will also be an aesthetic appeal to various shape combinations and uses of different shapes. Coverage can also be extended to include physics by discussing levers, forces resulting from levers, torques, and fulcrums.

COMMENTS: To extend this activity, a related topic for discussion is that of wind forces and how mobiles move. Also, as the mobile is being constructed, attention should be given to lengths of levers required for easy mobile motion.

*Mathematics Activities for Daily Living*

TEACHER COMMENTS TO STUDENTS:

Mobiles are relatively easy to construct and yet there is a lot of subtle mathematics and physics involved in the construction. Your class will have a mobile construction contest. This contest will have two basic objectives: (1) to provide materials that can be used to decorate the classroom; and (2) to generate competition to see who constructs aesthetically or mathematically designed mobiles.

*Judging rules*: evaluated on aesthetics; use of most geometric shapes.

These mobiles will be slightly different in that a cross piece must have a picture of an object on one end and a pipe cleaner model of that object on the other.

The picture (mounted on cardboard or something stiff) would be on one end of the wire and a pipe cleaner bicycle on the other, for example. It would look like Figure 5-3.

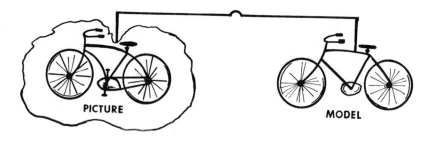

**FIGURE 5-3**

Use your imagination and see if you can create an unusual mobile using several different objects. Don't forget figures like stop signs.

Some of the models will be difficult to construct and might require some imagination to complete. For example, if you use a picture of a soft drink can, that shape (right circular cylinder) will be difficult to construct using a reasonable number of pipe cleaners. It could be shown as circles for the top and bottom, and a few cleaners joining the two circles.

*Mathematics Activities for Daily Living* 115

**TITLE:** **Draggin' On (5.08)**

GOAL: By considering the cost of building a drag strip, students will become aware of the subtle influences mathematics exerts within our society.

OBJECTIVES: The students will calculate partial costs related to building a drag strip.

GRADES: 7 - 12.

MATERIALS: Telephone, dimensions for drag strip.

INSTRUCTIONS FOR TEACHERS: A ticket to a drag strip will probably be at least four dollars. Using the cost as a generator, the investigation can be varied and extended, depending upon the age and interest of the class and the particular objective you have in mind.

For example, suppose you are doing a measurement unit and are discussing area and volume. The drag strip has five basic parts: the staging lanes (where cars are lined up to race, usually by class); the actual drag strip (where the cars are timed and their speed measured; the deceleration lane (slow-down distance after the race); return lane beside and parallel to the strip and deceleration lane); and pit area (where cars are repaired). The lengths and widths of these parts can be determined or given, and perimeters, surface area, volume and cost of paving material can be discussed.

If the concrete is the paving material to be used, the surface area will have to be converted to square yards since concrete is sold by the "yard" (cubic yard). It will be necessary to discuss thickness of the strip. Concrete four inches thick might crack because of weight or weather, so suppose the strip is six inches thick (use six inches thick because a "yard" of concrete would cover six square yards but eight inches of thickness could be used to complicate the situation). The students should call local surfacing suppliers to find just the cost of materials—concrete, for example. If they have computed the area needed, they could ask about a volume discount before figuring the cost. Retail prices of concrete will be such that the cost of materials alone for the drag strip will exceed $50,000. When other costs are considered, a four- or five-dollar admission ticket for the drag does not seem too high.

COMMENTS: This activity can be extended by discussing other costs in making a drag strip which would include land for the strip, stands and parking, restaurant facilities, rest rooms, timing devices, starting equipment, safety rails, stands and prizes as well as construction labor costs. Scaled drawings or models could also be made. You should determine the strip specifications before starting this (call local drag strip owner or write NHRA).[5] Perhaps you could get a strip owner to discuss with the class the mathematics involved. The student might be referred to "Speed Trap" for discussion related to how to find the dragster's top speed.

TEACHER COMMENTS TO STUDENTS:

Have you ever wondered why a ticket to the drag races is so expensive? This activity will help you answer that question by showing you part of the costs involved.

A drag strip has five basic parts as shown in the diagram in Figure 5-4. (This is not a scaled drawing):

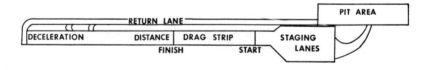

**FIGURE 5-4**

Suppose just the cost of paving material for these areas is considered. A few phone calls will provide you with information to answer the following questions:

---

[5]NHRA–Peterson Publishing Company, 8480 Sunset Boulevard, Los Angeles, California 90069

Length of drag strip  \_\_\_\_    Length of auxiliary
Width of each lane  \_\_\_\_        roads between strip
Length of timing at end              and return lane  \_\_\_\_
  (used to get top speed) \_\_\_\_  Width of auxiliary
Length of deceleration               roads between strip
  distance (allow enough             and return lane  \_\_\_\_
  for jet cars)  \_\_\_\_          Length of pit area  \_\_\_\_
Width of deceleration             Width of pit area  \_\_\_\_
  distance  \_\_\_\_                Length of staging
Length of return lane  \_\_\_\_      lane area  \_\_\_\_
Width of return lane  \_\_\_\_     Width of staging
                                     lane area  \_\_\_\_
                                  Number of staging lanes \_\_\_\_

With these specifications, you can calculate the surface area of each part which is to be paved.

Before computing the cost of paving the strip and related areas, you will need to consider necessary thickness of the parts concerned. For example, to avoid cracks or warps (which can be very dangerous at high speeds, and double-A fuelers run in the 230 mph range), the strip itself and the deceleration distance will need to be thicker. A call to a local builder would result in information which would tell how thick the parts would have to be.

Thickness of strip  \_\_\_\_       Thickness of pit area  \_\_\_\_
Thickness of deceleration           (Is it all paved?)  \_\_\_\_
  land  \_\_\_\_                   Thickness of staging
Thickness of return lane  \_\_\_\_    lane  \_\_\_\_

Using the surface areas computed above and the thickness for each part, compute the volume of paving material necessary to adequately surface the drag strip area.

Call local concrete and asphalt companies for the cost per "yard" (cubic yard) of material. (See Figure 5-5.)

|  |  |  | Volume Discount | | |
|---|---|---|---|---|---|
| Company | Cost | Type Material | Y\_\_ | N\_\_ | %\_\_ |
| \_\_\_\_ | \_\_\_\_ | \_\_\_\_ | \_\_\_\_ | \_\_\_\_ | \_\_\_\_ |
| \_\_\_\_ | \_\_\_\_ | \_\_\_\_ | \_\_\_\_ | \_\_\_\_ | \_\_\_\_ |
| \_\_\_\_ | \_\_\_\_ | \_\_\_\_ | \_\_\_\_ | \_\_\_\_ | \_\_\_\_ |

**FIGURE 5-5**

Knowing the volume of materials needed and the cost per "yard," compute the cost of materials to make the strip.

| Concrete cost_____ |
| Asphalt cost_____ |

Considering just this cost, how many four-dollar tickets would the owner have to sell if he were to get his money back (excluding interest)? How many five-dollar tickets?

Remember, all you did was purchase materials. These materials must be put down properly to make the drag strip proper and safe. But that means there will be labor costs involved in building the strip. What other costs would be involved in getting a drag strip ready for operation?

Now can you see why a ticket to the drag is four or five dollars?

# 6

# Activities That Teach Accuracy in Measurement

The laboratory or activity approach to the instruction of mathematics is commonly used in teaching measurement. The teachers of mathematics often ask children to measure objects in the classroom, to use a stopwatch to time classroom events, to measure the temperature of water as it boils or freezes, or to measure angles in a series of polygons. However, new situations involving concepts of time, angle measure, linear measure, and other measurement topics are frequently needed to provide for individual differences among students. This chapter offers measurement activities for children of different interests and backgrounds.

Experienced teachers are well aware that in a class of 30 students there are many different areas of interest. Activities in this chapter will appeal to students who are interested in such diverse areas as geography, biology, home construction and kite flying. The creative teacher of mathematics can use these activities as models to create many more interesting laboratory exercises for pupils.

The ability and achievement levels of students are an important factor in the study of measurement. Measurement can be as simple as determining the length of an object to the nearest inch or as complex as determining the wave length of infrared light. This chapter provides activities for many ability levels and the activities can be adjusted to handle more.

| | |
|---|---|
| **TITLE:** | **Square-Off (6.01)** |
| GOAL: | This activity will help students to become aware of uses for mathematics in different jobs such as surveying or general contracting. |
| OBJECTIVE: | Given a compass, directions and distances, the student will lay out shapes and compute the area and perimeter of each. |
| GRADES: | Geometry, 7, 8, General Mathematics. |
| MATERIALS: | Linear measuring devices, magnetic compass, stakes, hammer, string, paper, pencil, clipboard, and student instruction sheet. |
| INSTRUCTIONS FOR TEACHERS: | To encourage the students' actually measuring the figure and laying it out, perhaps the instructions could be given one at a time. The students would be told to go in a given direction and to a certain distance, at which point they locate a stake and then receive their next set of instructions. This procedure would be duplicated until the students return to the starting point. Failure to finish at the starting point would show that errors had been made. |
| COMMENTS: | The degree of complexity of this activity can be readily altered to fit the needs of a group of students. Better students could be asked to plot a regular hexagon with a given side length. some students could be required to calculate the angle sizes, locate the center, compute the perpendicular distance from the center to a side (and then check it on the actual figure), and determine the perimeter and area of the figure. In computing the area, a discussion could evolve on the use of congruent triangles and how this information, along with the area of one of the triangles, could be used in calculating the area of the entire hexagon. |
| | Many schools have a transit and other surveying tools. These aids stimulate interest and increase the accuracy enough so that it would be worth the additional instructional time to use them if possible. Furthermore, these instruments appeal to the more mature or older students. |
| | Small groups can be used with this activity, and the |

students should be encouraged to rotate job assignments as the activity progresses.

TEACHER COMMENTS TO STUDENTS:

Builders spend a lot of time staking out the area for a building. They need to know how long each side of the building will be as well as the number of corners so they can build according to plans.

In this activity you will be given a series of directions and distances, and you will use those pieces of information to lay out the described shape. Care should be taken to measure the angles and distances as accurately as possible because several small errors combined can make quite a difference in the final result.

Suppose you are building a home and you know where one corner of the house is to be located. Start at that corner and lay out the outside edge of the house according to the following instructions:

1. Go east 15 meters and drive a stake.
2. Go north 10 meters and drive a stake.
3. Go east 10 meters and drive a stake.
4. Go north 5 meters and drive a stake.
5. Go west 25 meters and drive a stake.
6. Go south 15 meters and drive a stake.

After step 6, you should be at the first corner stake that you started with. A question often asked of a builder is the amount of floor space available. Calculate the floor space and the perimeter of this house.

## TITLE: Anyday Calendar (6.02)

GOAL: Wouldn't it be nice to have a calendar that would be valid forever? This activity provides that calendar while employing set operations.

OBJECTIVES: Using four cubes, the student will mark the faces of those cubes in such a manner that they can be arranged to show month, day of the week, and date of the month.

The student will, using flat patterns, make out of heavy paper the four cubes.

GRADES: 6 - 12.

| | |
|---|---|
| MATERIALS: | Manila folder paper, marking pencil, ruler, scissors, tape or glue. |
| INSTRUCTIONS FOR TEACHERS: | The questions used here are developmental and the students should be encouraged to do each of them before proceeding. |
| COMMENTS: | There are a variety of mathematics concepts that can be stressed here, and each of these could be used as a means of inserting the activity into a program. For example, if a unit is being developed on flat pattern paper folding, it could be applied in this activity. One related question might involve determining the location of tabs necessary to complete the cube and minimize the number of them. However, if a more permanent calendar is desired, standard cut wooden two by two's can be used as a basic stock and then cut them at two-inch intervals, making a two-by-two-by-two cube. |

If using paper cubes, it might be desirable to complete all questions before folding the cubes and then have the student mark the faces while the cube is still unfolded.

Require that the markings be oriented in a manner which would permit reading all the faces when the completed cubes are rotated in a prescribed manner.

This answer section includes the response to each question in the order in which they are asked (to avoid numbers in the questions):

two

Jan
Feb

six

seven

Sat/Sun

1, 2, 3, ..., 29, 30, 31
0, 1, 2, ..., 7, 8, 9
1, 2, 3
0, 1, 2, ..., 7, 8, 9
0, 1, 2, 3

        6 or 3 on each
        3, 4, 5, 6, 7, 8, 9
        Turn 6 upside down when you need 9.

TEACHER COMMENTS TO STUDENTS:

    The calendar you make here can be used to give the month, the day of the month, and the day of the week for every year.

    A flat pattern for a cube should be duplicated and enlarged on your manila paper so that there are four patterns for cubes. If you want to minimize the number of cuts you need to make, you might consider variations of patterns which will still yield a cube. If you do that, don't forget the gluing tabs.

    One and only one of the cubes is to be used to show the month. Since all the months are to be listed on one cube, how many months must be listed on each face?_____ How can the names to be placed on each face be arranged so that there is no confusion as to which month is the current month? _____

    One and only one cube is to be used for the days of the week. How many faces are there on a cube?_____How many days in a week?_____How can the excess day be listed on the cube? _____

    The remaining two cubes must be capable of being arranged to show all days of the month. What numbers make up the days of the month? _____ Looking at these days of the month, what digits are needed to express the one's place? _____ Similarly, what digits are needed to express the ten's place?_____ Although it is not commonly done, this calendar expresses all days of the month as two-digit numerals so the first day is shown as 01. This is done so that all four cubes will be used all the time. Thus, a zero must be added to the list of those needed to express the ten's place.

    What is the union of the set of digits needed to express the one's place and the set of digits needed to express the ten's place?
_____

What is the intersection of the two sets of digits listed above? _____This intersection will be of elements which might occur twice, but since there is no 33rd day of the month, there is no need for two threes. Thus the three can be eliminated from this new set, originally determined by intersection.

You might reason in the same manner that there is no need to have two zeros. However, recall that the first nine days will be listed with a zero in the ten's place and therefore, a zero will be needed on each cube.

Recall that a zero, a one and two must appear on each cube in order to express the first nine days, eleven and twenty-two. How many faces will be left after putting a zero, a one and a two on each cube?_____List the digits that still must be placed on the remaining faces._____As you see, you have one more digit than face. How can this situation be handled?

---

Check to be sure all days can be listed with these two cubes.

## TITLE: Shape Up I (6.03)

| | |
|---|---|
| GOAL: | Students who intend further study in mathematics should be able to circumscribe regular polygons about a circle of a given radius. |
| OBJECTIVE: | The student will circumscribe regular polygons about a circle of given radius. |
| GRADES: | Geometry, Analytic Geometry, Algebra I, General Mathematics. |
| MATERIALS: | Compass, pencil and paper. |
| INSTRUCTIONS FOR TEACHERS: | Students should be encouraged to make generalizations; they can do it here and check their results. Generalizations like: "As the number of sides increases, the length of each side decreases"; and "The sides of the polygon being tangent to the circle necessitate that this tangent must be perpendicular to the radius." |
| COMMENTS: | See Shape Up II, the Comments section, for a discussion on how to combine Shape Up I and Shape Up II. |

TEACHER COMMENTS TO STUDENTS:

Occasionally regular polygons are circumscribed about a circle of given radius. Suppose an equilateral triangle is to be circumscribed about a circle of given radius. Since it is known that the figure has three sides, the measure of the central angles can be determined by dividing 360° by three. Each side of these angles will need to be extended beyond the circle. Bisect each of the central angles and draw each angle bisector so it is a radius. Construct a perpendicular to each of these angle bisectors at the endpoint which lies on the circle. In each case the perpendicular to the angle bisectors should be extended until the original central angle sides are intersected.

In this manner, the equilateral triangle is circumscribed about the circle. Similar procedures can be used to circumscribe other regular polygons.

Pictorially, this procedure has been as shown in Figure 6-1.

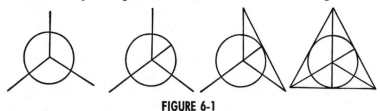

**FIGURE 6-1**

Is it possible to circumscribe an irregular polygon about a circle of given radius? Would the response to circumscribing an irregular polygon be altered if the polygon were either concave or convex?

**TITLE:** Shape Up II (6.04)

GOAL: Students should learn to inscribe regular shapes in a circle of given radius. This activity provides this desired awareness while blending geometry and algebra.

OBJECTIVE: The student will calculate the data necessary for inscribing a regular polygon in a circle of given radius.

GRADES: Geometry, Analytic Geometry, Algebra I, General Mathematics.

MATERIALS: Pencil, compass and paper.

INSTRUCTIONS FOR TEACHERS: The students should be encouraged to make generalizations and check their results. Encouragement should be given for students to make generalizations such as: "The central angle decreases as n, the number of sides, increases;" "If the radii of the circles remain constant, the length of the sides of the regular polygon decreases as n increases"; and "The measure of the central angle can be determined by dividing 360° by n."

COMMENTS: Attention should be focused on generalized statements involving fundamental algebraic manipulations to encourage the feeling that algebra and geometry can be studied jointly at times and that each area can be helpful and informative in the study of the other.

This activity, along with Shape Up I, has the student investigate inscribed and circumscribed regular polygons respectively, all on circles of a constant radius. Combining the two activities can develop important background for later material. This would refer most directly to the limit process employed in calculus where the width of a series of rectangles just inside and just including a given curve is decreased. As the width of the rectangle is decreased, the difference between the area above and below the curve decreases. A limit situation can be generated in this activity by inscribing a series of regular polygons in a circle of given radius.

If both activities are used concurrently, graph paper should be used to simplify comparisons by having similar polygons done on the same circle. The following paragraph should be added to the student's sheet:

After you have circumscribed each figure with its respective inscribed figure, visually compare the difference between the area of the circumscribed shape and that of the similar inscribed shape for each circle. Does there appear to be some general statement that can be made about the area difference as "n" increases? Use the graph paper to verify your conclusion.

TEACHER COMMENTS TO STUDENTS:

There are a variety of methods that can be used to inscribe regular polygons in a circle. One method entails making congruent central angles and joining the points determined by the intersection of the sides of these angles with the circle. For example, the inscribed equilateral triangle needs three central angles of 120°. Let the sides of these three adjacent central angles be radii of the circle. From the three points where the three radii intersect the circle, an equilateral triangle is formed. (See Figure 6-2.)

Given circles of radius r, inscribe one of the following regular polygons in each circle: square, pentagon, hexagon and heptagon. Can you make a general statement about the length of sides of the polygons, the size of the central angles, or how the central angle can be calculated for each n-gon?

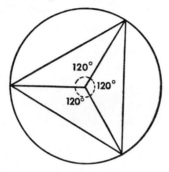

**FIGURE 6-2**

| | |
|---|---|
| **TITLE:** | **SymMETRIC (6.05)** |
| GOAL: | One objective of mathematics instruction is to develop an awareness in the student of mathematics in situations frequently taken for granted. Symmetry exists in many areas with which the student is quite familiar. These areas can be used to strengthen the understanding of the concept. |
| OBJECTIVE: | The student will determine lines of symmetry (if they exist) of shapes and letters. |
| GRADES: | Geometry, 7, 8, General Mathematics. |
| MATERIALS: | Printed alphabet (both capital and lower case letters), mirror, and regular polygons. |
| INSTRUCTIONS FOR TEACHERS: | Encourage the students to be flexible and open in their thinking, thus helping them to understand that they might need to view the letters and shapes from "non-standard" vantage points. When placing the mirror on the shape of letter, care must be taken that the plane of the mirror is perpendicular to the |

|  | plane of the paper on which the shape or letter is printed. |
|---|---|
| COMMENTS: | More extensive activities such as these are available in texts or in the literature. It is relatively easy to generate similar situations within the classroom. |

TEACHER COMMENTS TO STUDENTS:

Symmetry lines are found in shapes and letters as well as in many other things. An equilateral triangle, for example, has three lines of symmetry (Figure 6-3).

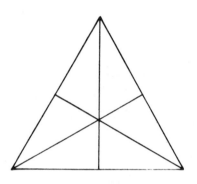

**FIGURE 6-3**

However, an isosceles triangle has only one line of symmetry—the line bisecting the angle formed by the two congruent sides. Would a scalene triangle have a line of symmetry? Use your mirror to check the results. Can you make a general statement about the lines of symmetry for triangles?

Now check each member of the quadrilateral family for lines of symmetry, beginning with squares. Can you make a general statement about the lines of symmetry for quadrilaterals? How does this general statement compare with the one for triangles?

Other polygons have lines of symmetry. Considering only regular polygons, can you make a general statement about the numbers of lines of symmetry? How many lines of symmetry does a circle have?

Symmetry lines occur in letters of our alphabet. "A" is symmetric about a vertical line bisecting the angle formed by the two long sides while "D" is symmetric about a horizontal line from a

*Activities That Teach Accuracy in Measurement*

**FIGURE 6-4**

point halfway up, or down, the straight side. "I" has two lines of symmetry, one vertical and one horizontal. What other capital letters have lines of symmetry. Are they horizontal, vertical or oblique? Do any of the small case letters have lines of symmetry?

What word is this? Put your mirror on the dotted line to find out. (See Figure 6-4.) Make up some "half words" of your own.

Identify and describe lines of symmetry for the digits 0 through 9. Can you combine digits to form numerals which have vertical lines of symmetry? Is there any multi-digit numeral that has a horizontal line of symmetry?

**TITLE:** Peel It! (6.06)

GOAL: The study of projections of a global map onto a plane is a difficult concept for students to understand. This activity is designed to give a concrete experience with this difficult concept.

OBJECTIVE: The students will make a plane representation of a spherical net.

MATERIALS: One orange per student, a Magic Marker and kitchen knife for every two or three students, at least one globe.

GRADES: 6, 7, 8, General Mathematics, Geometry.

INSTRUCTIONS FOR TEACHERS: Separate the class into pairs of students or groups of three, and give careful directions about the proper behavior with knives.

Allow the students sufficient time to explore the lines of latitude and longitude on the globe. You might ask questions about the location of various geographical points of interest, using approximate longitude and latitude.

When the students begin to work with the oranges, be certain that they draw lines on the orange where they will make cuts. Drawing the lines allows the small group to discuss the placement of the lines before cutting. Haphazard cutting without lines often produces very unsatisfactory results and arguments among students.

COMMENTS: This is only an introductory activity. You may wish to consider this problem further with the class. Library research about different types of projections is often rewarding to the geometry student.

TEACHER COMMENTS TO STUDENTS:

Flat maps of the world are often made in weird shapes. Why is this so?

Study your globe and notice the latitudes and longtitudes. The latitudes are the parallel lines on the globe. The equator is one of these lines. The lines which pass through the north and south poles are lines of longitude. Do you see a way that the surface of the globe could be drawn on a flat surface?

Let's look at another sphere like the globe and try to find the answer to our problem. Use an orange as our sphere. Peel the skin off and try to flatten it. Be careful that the skin remains in one piece!

To aid the peeling of the orange, draw lines on the surface where you want to make a cut in the skin.

For example, see Figure 6-5.

The dotted lines form a spiral. What would it look like as a map?_____

**FIGURE 6-5**

A map as long and skinny as the one produced by this example would not be practical. Cut another orange to produce a better map which is flat!

Be sure to draw lines on the orange where you will cut the orange!

Draw a picture of your orange peel flattened out. Would it make a good map? _____

## TITLE: Paper Protractor (6.07)

GOAL: Through the student's knowledge of symmetry, a protractor is made in this activity.

OBJECTIVES: The student will construct a circle.
The student will fold a circular piece of paper to form angles on a protractor.
The student will measure angles with a protractor.

GRADES: 6, 7, 8, General Mathematics, Geometry.

MATERIALS: Compass, tracing paper, scissors.

INSTRUCTIONS FOR TEACHERS: These questions are developmental, and therefore the student should answer each one before going on.

COMMENTS: Many students have difficulty reading a protractor which is scaled from both right to left and left to right. In making these paper protractors the students may make two—one marked for clockwise measurement and the other for counterclockwise measurement.

TEACHER COMMENTS TO STUDENTS:

Use a compass to construct a circle with a 20 centimeter diameter on the tracing paper. Cut out the circle.

Now fold the circle in half very carefully; cut the halves apart.

Mark the center of the diameter of the semi-circle clearly with a pen or pencil mark. (See Figure 6-6.)

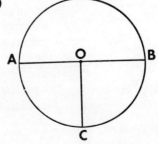

**FIGURE 6-6**

What is the measure of angle AOB?

Yes, since angle AOB = 180°, we can now write our first angle measurements.

What is the measure of angle AOC? Write 90° at point C.

Fold the semi-circle to find one-half of the 90° angles.

Label these angles!

Now, to find other convenient angles between the 0° and 45°, the 45° and 90° and 135° and 180° marks, the angles must be folded into thirds.

To accomplish this, fold each angle much as a business letter is folded, into thirds. (See Figure 6-7.)

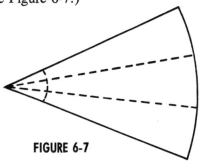

**FIGURE 6-7**

What are the names of the new angles?

Label your protractor.

You now have a protractor labeled with angles of 0°, 15°, 30°, 45°, 60°, 75°, 90°, 105°, 120°, 135°, 150°, 165° and 180° in a clockwise manner. However, not all angles can be conveniently measured in a clockwise manner. So use the second semi-circle to make a protractor in a counterclockwise manner.

Once you have made your second protractor, measure some angles!

### TITLE: Make a Kite! (6.08)

GOAL: We need to develop students' awareness of the use of mathematics in their daily activities. Many times the students just don't think of mathematics as being used anywhere outside the confines of the classroom.

OBJECTIVE: The student will, while making a kite, measure various lengths.

| | |
|---|---|
| MATERIALS: | Two sticks either 1 cm square or 1/2 cm by 1-1/2 cm, tissue wrapping paper, string, white glue, tape, and scissors. |
| GRADES: | 6, 7, 8. |
| INSTRUCTIONS FOR TEACHERS: | As the kite is being constructed, only small amounts of glue should be used because excessive amounts will tear the paper and add weight. Also, the glue should be placed at the edge of the paper to avoid a flapping effect from unglued portions while the kite is in flight.
The bridle of the kite, which is attached to the upright or longer stick, is crucial to good flight. The lengths of the bridle segments may have to be altered for better results.
One variation which might be considered is the insertion of a rubber band into each segment of the bridle. The rubber bands will allow the kite to adjust itself to the different wind currents. |
| COMMENTS: | Although the instructions here are for a standard bow kite, many other types of kites can be made. It is conceivable that interest could be developed to such an extent that kites of exotic design could be constructed and the geometry involved pursued. Even bow kites could be decorated with assorted designs, thus creating another means of discussing geometry. |

TEACHER COMMENTS TO STUDENTS:

In making this kite, the longer of the two sticks is the upright, the one to which the tail is tied. The shorter stick is the crosspiece and is the one which will be bowed eventually. Both sticks must be notched at each end so that a string may be fitted into that notch (Figure 6-8).

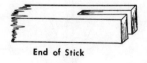

End of Stick

**FIGURE 6-8**

The two sticks should be lashed together to form a cross. The shorter stick should be lashed at its center and the longer should be lashed about 2 decimeters from the top (Figure 6-9).

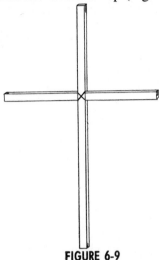

**FIGURE 6-9**

Before papering the kite, a string loop must be placed around the outside ends of each stick; from the top to the ends of the crosspiece the string should be tight, while from the ends of the crosspiece to the bottom the string should be a little slack (Figure 6-10). Tie the string. With the bottom string having a little slack, the kite will fly in a more stable manner.

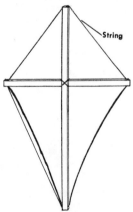

**FIGURE 6-10**

*Activities That Teach Accuracy in Measurement*　　　　　　　　　*135*

A variety of paper can be used, from light tissue to heavier newsprint or wrapping paper. Lighter paper gives a lighter kite but it is also more fragile. Although sheets of paper can be glued together to make a covering, it is better to have one large sheet.

Place the paper on a smooth flat surface with the side that is to be the front of the kite underneath. Place the stick and string frame on the paper, cutting the paper 2 cm. outside the string of the kite. The dashed lines in Figure 6-11 show where the paper should be cut.

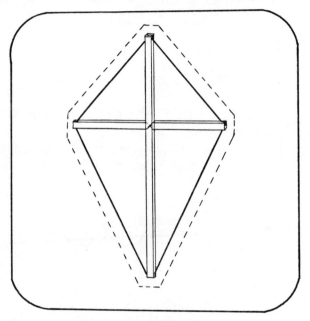

**FIGURE 6-11**

You are now ready to glue the paper over the boundary strings. Cut the paper into the string about 5 mm. down from where the string meets the stick, making the cut perpendicular to the string. *Do not cut the string!* You should make eights cuts, two at each of the four ends of the sticks. Using only a very small amount of glue and placing it right at the cut edge, put the glue along one of the edges from one perpendicular cut to another (Figure 6-12). Do not glue the little paper tab that would be at the ends of the sticks.

Fold the glued edge over the string and press it down. The glue is placed at the edge of the paper so that wind can't catch and tear it.

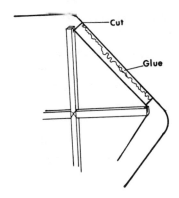

**FIGURE 6-12**

Therefore, if any of the edge sticks up, glue it down. Glue the remaining three edges in the same manner.

Each stick end has a paper tab over it which should be cut off so that the stick end and some string show from the front. (See Figure 6-13.)

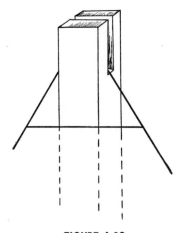

**FIGURE 6-13**

In bowing a kite, the crosspiece is bent and a string is tied to each crosspiece end to hold the bend. Looking at the kite from the top, the bow string should be about 12 cm. above the point where the sticks cross. (See Figure 6-14.)

**FIGURE 6-14**

The bridle is attached to the longer (unbowed) stick, one decimeter from the top and half the kite length (half of 9 dm.) from where the top bridle string is attached. After locating where the bridle string is to be attached, put a piece of tape over the points. Make a small hole in the tape and paper, pass one end of the string through the hole, and tie it securely to the upright stick. (See Figure 6-15.)

The bridle string should be about a meter long. Attach the flying cord to the bridle and fasten it tightly. Sliding the flying cord up or down the bridle will change the angle of flight. You are now ready for a test flight. Do you need a tail for the kite?

A kite can be made larger or smaller than the one described as long as the general ratios are not changed too much.

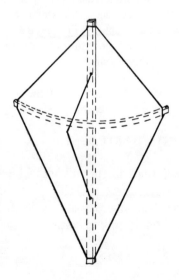

**FIGURE 6-15**

| | |
|---|---|
| **TITLE:** | **Kite Pulling Contest (6.09)** |
| GOAL: | The student will design and construct a kite(s) following given guidelines, and compete in a kite "tug of war." |
| OBJECTIVES: | The student will design a kite using his knowledge of geometry, and he will use measurement techniques to insure that his kite has a surface area within a stated maximum. |
| | The student will, through trial and error and knowledge of vectors, determine the optimal height and angle for flying a kite in a "tug of war" contest. |
| GRADES: | 7 - 12. (The sophistication of kite design, the precision of measurement, and the use of vectors, all allow for adaption to different grade levels.) |
| MATERIALS: | A strong paper (meat wrapping paper), glue, string, scissors, sticks, two long stakes (at least four feet long), and two pulleys. |
| INSTRUCTIONS FOR TEACHERS: | 1. Reproduce the directions to students and distribute a copy to each student. |
| | 2. Encourage each student to build a kite. If teams are necessary because of a lack of materials or student apathy, limiting the teams to two members will aid in insuring a desirable level of student involvement. |
| | 3. You may wish to involve the industrial arts (construction ideas) and/or the physics teacher (vectors) in a team teaching venture. |

TEACHER COMMENTS TO STUDENTS:

The mathematics department will hold a kite pulling contest on _____(date) at_____(place). The final eliminations will be held on that same day after school. Each mathematics student will have an opportunity to compete in his own mathematics class.

*Building of Kites:*
1. You may choose any design that you wish:
    a. a hexagon with three sticks,

b. a traditional two-stick kite, or
c. a delta wing kite.
2. However, you must conform to the following rules:
   a. The kite must have only one pulling surface. (This eliminates box kites.)
   b. The surface area of the pulling side of the kite must have no more than 2500 square inches.

*Flying of Kites:*
1. You may fly your kite at any altitude. Your kite will be tied to one other kite for each pull. Figure 6-16 shows you how this is to be done.

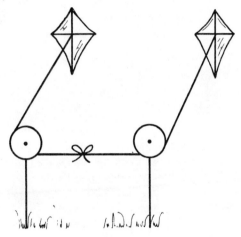

**FIGURE 6-16**

2. After the kites are tied, you must release your kite. The kite which pulls the connecting knot through its pulley is the winner (Figure 6-17).

**TITLE:**     **Boxed In (6.10)**

GOAL:     Occasionally it is advantageous for the teacher to become involved in activities as a participant. The students are highly motivated by competing in an activity in which they compare the volumes of similar prisms.

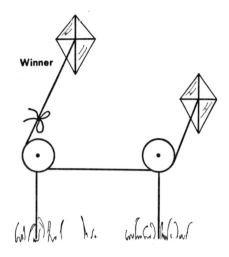

**FIGURE 6-17**

OBJECTIVES: Given construction paper and specific dimensions, the student will construct a rectangularly based right prism. The student will compute the number of smaller boxes all of the same dimension which will fit in a larger box of specific dimensions.

MATERIALS: Construction paper, scissors, tape or glue.

GRADES: 6, 7, 8, General Mathematics.

INSTRUCTIONS FOR TEACHERS: The students need to be reasonably accurate in their construction because they will be placing their boxes into the one you make. They will see the number of smaller boxes it takes to fill a larger one and should become more aware how increases in dimension affect volume. Your box should have dimensions of 4 × 6 × 5 units.

COMMENTS: This activity can involve a large number of students if the dimensions of your box are suitable multiples of the dimensions of the students' boxes.

The degree of complexity can be reduced by working with cubes. Usually the students see the change in volume more readily.

Whether you use cubes or rectangular based right prisms, it might be advantageous to make your box out of plexiglass. With this, the students could actually see how many of the small boxes fit in the large one.

A student could be assigned the responsibility of making the large box.

TEACHER COMMENTS TO STUDENTS:

You are to make a box out of construction paper. The dimensions are to be 1 × 2 × 2-1/2 units as shown in Figure 6-18.

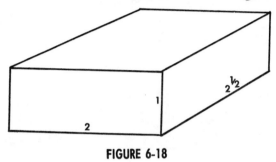

**FIGURE 6-18**

There will be a larger box of the following dimensions—4 × 6 × 5 units. How many boxes like the one you make will fit into the larger box?

When making your box be careful to be as accurate as possible. Yours along with others will be placed inside the larger box.

Before placing the little boxes in the big box, estimate how many of the smaller ones it will take to fill the larger one to capacity. Now, check your estimate. How close were you? Is there a better way to find how many small boxes would fit in the large one?

To find the volume of a box you multiply the three dimensions, in your case 1 × 2 × 2-1/2 units, and the product will be in cubic units. This answer tells how many one-unit cubes would fill this box (even though some of them might have to be cut up). What is the volume of your box?_____

What is the volume of the large box if the dimensions are 4 × 5 × 6 units?_____

Divide the volume of the small box into the volume of the large box. What did you get?_____Doesn't that answer look familiar? It should. It is the same as the number of small boxes that fit into the large box. This will always work.

# 7
# Simplified, Interesting Ways to Teach Geometry

The study of geometry in public schools has traditionally been limited to the study of plane and space geometry with Euclidean restrictions. However, teachers of mathematics who are concerned about providing for individual differences among students have found that the techniques of Euclidean geometry are not sufficient to teach geometric concepts to all students. This chapter includes activities involving analytical geometry, projective geometry, transformational geometry, and Euclidean geometry to provide the teacher samples of a variety of approaches for teaching geometry through activities.

Experienced teachers are well aware that students often have difficulty grasping abstractions in geometry. Thus, students must be provided with experiences on the concrete level. Activities in this chapter, such as "String It," provide teachers with activities which require students to use manipulatives to explore the nature of abstract geometrical concepts.

Students can be motivated to study geometry through activities, also. "Dunebuggy Race" and "Soap Bubbles" exemplify this purpose. The creative teacher will be able to generate similar activities based upon a particular student's interest. Puzzle activities often motivate students to study a concept with interest for long periods of time. "It's Obvious, Isn't It?" is an example of only one of many puzzles based on geometric concepts that the teacher can utilize to motivate students.

| | |
|---|---|
| **TITLE:** | **String It (7.01)** |
| GOAL: | The students will explore the highly abstract topic of invariance of properties in projective geometry by using concrete materials. |
| OBJECTIVES: | The student will sketch the projected images of geometric figures in concrete situations and in paper-pencil exercises. |
| | The student will state whether or not certain properties are invariant under given projections. |
| GRADE: | Geometry. |
| MATERIALS: | Ring stand, thread, ½" × 10" × 2' pine boards, saw, poster board, thumbtacks. |
| INSTRUCTIONS FOR TEACHERS: | Prepare the board (½" × 10" × 24") with a ⅛" saw cut, ¼" deep, 8" from the end. Center the ring stand at the end of the board nearest the cut and attach 12 strings each 26" long to one point on the ring stand. Prepare cardboards with geometric cut-outs of a circle, ellipse, square, and line segments that would intersect if extended. |
| | When the student tacks the string to the board, he should make sure that the string forms a *straight* line with the point on the cut-out and the point where the strings are tied (the focal point); otherwise he could make any kind of figure. |
| COMMENTS: | For an added problem, the students should place a paper on the board, draw a figure, tack the strings to the figure, then decide what geometric figure would be needed to project the figure drawn. This can be done by cutting a piece of paper until it fits the strings at the place where the perpendicular card would be placed. Also, an activity which uses an actual light and transparencies can be done in conjunction with this activity. |
| | Some students may wish to explore the projection concepts inherent in cartography. For example, Mercator projections can be simulated with a transparent globe lighted internally and a cylinder of paper. |

*Simplified, Interesting Ways to Teach Geometry* 145

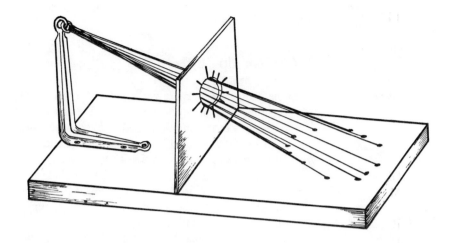

**FIGURE 7-1**

TEACHER COMMENTS TO STUDENTS:

Projections in geometry are sometimes hard for students to visualize. This project should help you form a mental model of what happens to points which are projected from a focal point through a figure to a plane perpendicular to the plane of the figure. Imagine the point where the strings are tied to be a light source and the strings as rays of light.

(1) Insert a card containing a cut-out of a circle in the cut in the board.
(2) Take each string and pass it through the cut-out, touching the circle at the points marked. Hold the string on the board, move it in and out, left and right until it forms a straight line with the point where the strings are tied and the point on the cut-out. When you are sure the line is straight, tack it to the board. Do this for each of the 12 strings. What figure is mapped by the tacks?
(3) Insert the ellipse card and perform the same operations. What figure is mapped by the tacks?

(4) What figure is mapped by the tacks when you use the square cut-out? The _____ cut-out?

**TITLE:**     **Rip 'em Up (7.02)**

GOAL: Some geometric facts are verifiable with certain procedures used on physical materials. The sum of the measures of the angles of a triangle equaling 180° can be demonstrated by cutting the triangle into three pieces, each of which contains one of the triangle's vertices. If the angles of these three vertices are placed so that they are adjacent, the exterior edges of the three will form a straight angle, the measure of which is 180°.

OBJECTIVE: The student will develop a formula for finding the sum of the interior angles of a polygon.

GRADES: Geometry, 7, 8, General Mathematics.

MATERIALS: Construction paper or cardboard tablet back, scissors, straight-edge and pencil.

INSTRUCTIONS FOR TEACHERS: Care must be taken to insure that each student understands the procedure to be followed as a polygon is cut and the vertices placed together. Each student's work should be checked initially with the triangle and then again, perhaps when he gets to the pentagon, since it is the first situation requiring an overlapping.

Notice that there are three different places where the student is asked to state a generalization. If a student cannot state the desired conclusion when first requested, additional information is supplied and eventually the request is repeated. Each student should be encouraged to state the generalization as soon as possible, thus enhancing his ability to formulate hypotheses. The additional information is intended for those who have difficulty reaching conclusions at abstract stages of development.

COMMENTS: This activity could readily be used in a number theory section of an algebra class; wherever it is used, the blending of algebra and geometry should be stressed.

# Simplified, Interesting Ways to Teach Geometry

TEACHER COMMENTS TO STUDENTS:

Draw a triangle on your construction paper.
Cut it out.
Label each angle of the triangle.
Cut the triangle into three pieces so that each piece has one angle from the original triangle.

Place the three original angles so that their vertices are a common point and the sides of the middle piece touch a side of each of the outside pieces with no overlapping.

Pictorially, the procedure has been as illustrated in Figure 7-2.

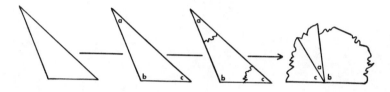

**FIGURE 7-2**

What is true of the final figure which has been constructed? How many degrees are in the angle formed by the outside edges of the two outside pieces? Record your results in Table 1. Would this result be true for all triangles?

Draw a quadrilateral on your construction paper.

Label each angle of the quaderilateral. (For this discussion, use angles a, b, c and d.)

Cut the quadrilateral into four pieces so that each piece has one angle from the original quadrilateral.

Place angle "a" on a surface. Put angle "b" beside angle "a" so that they have a common vertex and their sides are touching. Now place angle "c" with angles "a" and "b" so that all three have a common vertex and also place it so that a side of angle "c" touches a side of either angle "a" or angle "b". Angle "d" should fit in the remaining space so that it has the common vertex that the other three angles have and the sides of angle "d" touch the sides of two of the three angles with no overlapping.

Pictorially, the procedure has been as shown in Figure 7-3.

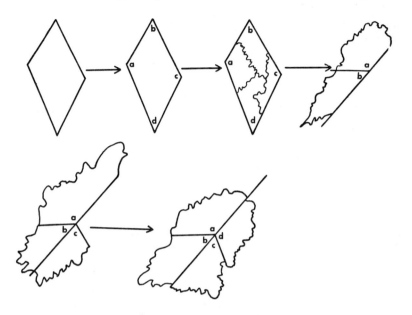

**FIGURE 7-3**

What is the measure of the angle formed by the outside edges of the first and last pieces? Will this be true for all quadrilaterals? Record the results on a table, as shown in Figure 7-4.

| Number of Sides of the Polygon | Number of Vertices of the Polygon | Measure of Angle Formed by Outside Edges of the First and Last Pieces |
|---|---|---|
| 3 | 3 | 180° |
| 4 |   |   |
| 5 |   |   |
| 6 |   |   |
| 7 |   |   |

**Table 1**

**FIGURE 7-4**

Repeat the procedure for a pentagon. Notice that the resultant measure will require an overlapping of the first piece(s) by the last piece(s).

Pictorially, the procedure should be as pictured in Figure 7-5.

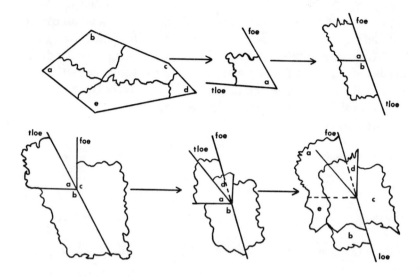

foe = first outside edge
tloe = temporary last outside edge
loe = last outside edge

**FIGURE 7-5**

The resultant angle would have the final pieces overlapping the first pieces and the angle measure would be that of one and a half circles.

Repeat the procedure for a hexagon.

Repeat the procedure for a heptagon.

Can you state a generalization for finding the measure of the sum of the interior angles of an n-sided polygon? If so, do it; if not, proceed.

For the triangle, the sum of the measures of the angles was 180°, which is a semi-circle or $\frac{1(360°)}{2}$. The sum of the measures of the angles of the quadrilateral was 360° or $\frac{2(360°)}{2}$. The pentagon gave a sum of 540° or $\frac{3(360°)}{2}$. The hexagon gave a measure sum of 720° or $\frac{4(360°)}{2}$ and the heptagon gave 900° or $\frac{5(360°)}{2}$.

Express each of the sum measures as the product of two factors, one of which is 180°. In each of these, subtract the factor which is not 180° from the number of sides of the polygon used to get that particular measure sum. What is the difference in each case?

Can the number of sides of the polygon, along with the differences calculated in the last paragraph, be used to express the final sum of the measures of the interior angles of the polygons?

Can this be used to state a generalization?

If not, consider:

For the triangle $\quad 180° = \dfrac{1(360°)}{2} = \dfrac{(3-2)(360°)}{2}$

For the quadrilateral $\quad 360° = \dfrac{2(360°)}{2} = \dfrac{(4-2)(360°)}{2}$

For the pentagon $\quad 540° = \dfrac{3(360°)}{2} = \dfrac{(5-2)(360°)}{2}$

For the hexagon $\quad 720° = \dfrac{4(360°)}{2} = \dfrac{(6-2)(360°)}{2}$

For the heptagon $\quad 900° = \dfrac{5(360°)}{2} = \dfrac{(7-2)(360°)}{2}$

Notice that in each case, the number of sides minus two is multiplied by 360° and that product is divided by two.

In general then, what is a formula for finding the sum of the measures of the interior angles of an n-sided polygon?

## TITLE: It's Obvious, Isn't It? (7.03)

GOAL: This activity will afford the student the opportunity to explore the nature of geometric shapes and some apparent dilemmas caused by manipulating them.

OBJECTIVES: The student will draw given geometric figures.

The student will measure distances on geometric figures.

The students will manipulate given geometric figures to form other geometric figures.

MATERIALS: Construction paper, scissors.

GRADES: 8, General Mathematics, Geometry, Trigonometry.

INSTRUCTIONS FOR TEACHERS: The parts of the first triangle may be hinged to form a square, as shown in Figure 7-6.

In the second activity, the student instructions leave the rigor of the proof that $64 \neq 65$ to the level of the student.

Simplified, Interesting Ways to Teach Geometry           151

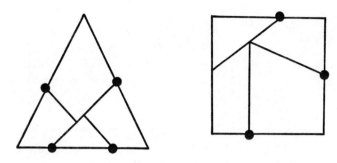

**FIGURE 7-6**

TEACHER COMMENTS TO STUDENT:

1. Draw an equilateral triangle with sides of eight inches on the construction paper.

Now, measure carefully and mark the indicated points on the triangle. Cut the triangle on the dotted lines, but be sure that the two indicated dotted lines form right angles! (See Figure 7-7.)

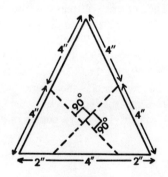

**FIGURE 7-7**

Can you arrange the pieces to form a square? Try it!

2. Draw a square with eight-inch sides on the construction paper (Figure 7-8).

What is the area of this square?_____

Now cut the square as shown by the dotted lines (Figure 7-9).

Form a triangle with the pieces.

152    *Simplified, Interesting Ways to Teach Geometry*

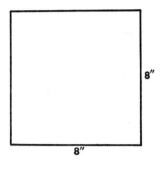

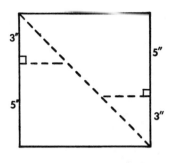

**FIGURE 7-8**              **FIGURE 7-9**

What is the area of the triangle? _____

How do you explain that the area of the square is 64 inches, but the area of the triangle is 65 square inches?

Check the first example! Are the areas equal? If they are not equal, why not?

| | |
|---|---|
| **TITLE:** | **Dunebuggy Race (7.04)** |
| GOAL: | The student will find the "shortest course" on a given race track. |
| OBJECTIVES: | The student will measure the distance around a race track with a trundle wheel. |
| | The student will verbalize whether or not a race track can be traveled without retracing a section. |
| GRADES: | 7, 8, 9, 10, 11, 12. |
| MATERIALS: | Trundle wheels, shelf paper or rolls of newsprint. |
| INSTRUCTIONS FOR TEACHERS: | 1. Draw the race track shown in Figure 7-10 on shelf paper. Ten or 12 copies of each will be required. |

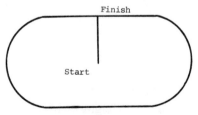

**Track 1**

**FIGURE 7-10**

*Simplified, Interesting Ways to Teach Geometry* 153

      2. Group the students into driver-navigator teams.
      3. Discuss the types of race tracks that can be traversed without retracing, by describing the tracks according to their vertices. An even vertex has an even number of roads stemming from its junction. Similarly, an odd vertex has an odd number of roads leading from it.
      4. Euler described four generalizations about networks or race tracks of this type. A good reference to these generalizations and a review of networks is: Johnson, D.A. and N. H. Glenn. *Topology: The Rubber Sheet Geometry.* New York: McGraw-Hill, 1960, pp. 14-18.

COMMENTS:   If racing seems to motivate your students highly, consider using this activity along with "Slotcar Racing."[5]

TEACHER COMMENTS TO STUDENTS:

Today we are going to look at dunebuggy racing. The Metropolitan Racing Association is sponsoring a race in which the rules are:

1. Drive your vehicle following the posted speed limits exactly.
2. Only one vehicle is permitted on the track at any one time.
3. Every race consists of one complete lap. Every section of the track must be traveled at least once.

How does a team win? Simple! The driver and navigator must find the shortest lap on the track.

O. K.—racing team, choose race track number 1 from the pit area and a trundle wheel to measure the track. Track 1 should look like the track drawn on the chalkboard (Figure 7-10).

What is the shortest distance around this track?_____
Choose Track 2 (Figure 7-11) from the pit area. Select a starting point! What is the distance around this track?_____
If you retraced any portion of the track, you have not found the

---

[5]Brumbaugh, D.K., and Hynes, M.C., "Slotcar Racing," *School Science and Mathematics* (74), May-June, 1974, pp. 447-48.

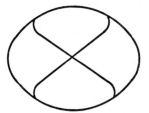
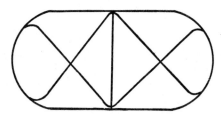

Track 2            Track 3

**FIGURE 7-11**         **FIGURE 7-12**

shortest path. Try again! Can you travel around the track a different way from the same starting point?_____Try a different starting point. Can you travel a different way?_____

What is the shortest distance around Track 3 (Figure 7-12)?
_____

Did you have to travel any section of the track twice?_____

Now, it's your turn to draw a race track. Draw one which cannot be traveled without retracing and challenge the other racing teams to find the shortest path.

## TITLE: Soap Bubbles (7.05)

| | |
|---|---|
| GOAL: | This activity will use a hobby to allow students the opportunity of exploring the world of geometry. |
| OBJECTIVES: | The student will build geometric space figures. |
| | The student will state the space figures produced by soap bubbles in the constructed space figures. |
| MATERIALS: | Choice of sets: |

                       1. coat hanger wire—brazier  
                       2. copper wire—solder       } and liquid soap  
                       3. aluminum wire—epoxy  
              or 4. plastic sticks and liquid plastic.

Needle nose pliers are helpful if wire is used.

| | |
|---|---|
| GRADES: | 5, 6, 7, 8, General Mathematics, Geometry. |
| INSTRUCTIONS FOR TEACHERS: | Choose the materials for this activity according to the level of the class. For example, in a general mathematics class with many boys enrolled in a metal shop, copper wire and soldering or coat hanger wire and brazing may be the most desirable. However, for younger students the plastic or light wire and glue may be most appropriate. |

*Simplified, Interesting Ways to Teach Geometry* 155

                    Since this activity does require special equipment in some instances, this may become a team-teaching effort with the industrial arts and/or art teachers.

                    When the students are making the frames, encourage them to bend the wire to form corners wherever possible. Every joint must be firmly secured to insure that the soap film will attach to the joint later.

                    The students may be more successful in this endeavor if they have studied Euler's "rules" of tracing networks.

                    Assign students to work in pairs in this activity since an extra pair of hands is often needed to secure the joints of the frames.

COMMENTS:    If some of the students become interested in this activity encourage them to read *Soap Bubbles: Their Colours and the Forces Which Mould Them*, by C. V. Boys, available from Dover Publications. You may also want to make a Mobius bubble frame since the Mobius strip is often highly motivating for students.

TEACHER COMMENTS TO STUDENTS:

Using your framing materials, make the geometric figures shown in Figure 7-13.

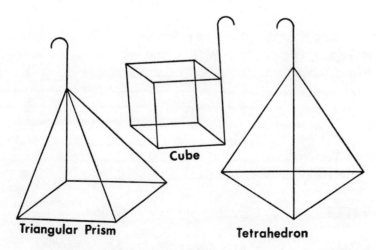

**FIGURE 7-13**

As you make the frames, it is best not to cut the wire for each edge of the figure. Bend the framing material when you can since this reduces the number of joints that you must make. Make sure that all joints are securely made!!

Now, hold one of your frames by the handle and dip it in the soap solution. Is the result what you expected? Before you try the second frame, try to predict how the "bubble" will look.

Using the cube, try to get the result shown in Figure 7-14.

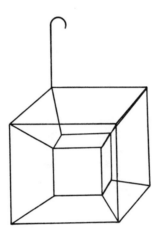

**FIGURE 7-14**

This is called a hypercube or a 4th dimensional cube. Try to make a hypertetrahedron! What do you notice about the faces of the hypertetrahedron? _____
_____

Write down some of your observations about the bubbles.
Cube: _____
Tetrahedron: _____
Triangular prism: _____

**TITLE:**    **Circular Shadows? (7.06)**

GOAL:    Often a sunny spring day is not conducive to drill and practice lessons. This activity provides a worthwhile geometry lesson in the sun.

*Simplified, Interesting Ways to Teach Geometry* 157

OBJECTIVES: The student will construct a circle of given diameter.
The student will draw a rectangle of given dimensions.
The student will draw shadows of circles and rectangles projected on a plane surface.

MATERIALS: Construction paper, scissors, paper soda straws, glue.

GRADES: 6, 7, 8, General Mathematics.

INSTRUCTIONS FOR TEACHERS: Do this activity only on a sunny day! Be sure that the circle and rectangle are not taken outside at the same time. The rectangle activities provide opportunities for the students to make inferences. An important part of this activity is sharing drawings and discussing the results of Part I so that better inferences may be drawn about the shadows of the rectangle.

COMMENTS: The straw may be attached to the rectangle on the width or the length. Thus, there will be varying answers in Part II.

TEACHER COMMENTS TO STUDENTS:

**FIGURE 7-15**

I. Use your compass to draw a circle on construction paper. Make the circle have a 4-inch diameter. Now, cut the circle from the paper, and glue a straw to the shape so that it looks like a lollipop (Figure 7-15). Take this device outside and sit on the sidewalk. Hold the circle parallel to the ground. How does the shadow look? Is it a circle?_____ Draw a picture of this shadow and label it "parallel." Hold the shape perpendicular to the ground. Draw this shadow, and label it "perpendicular."

Move your circle around. Can you find a position that makes a circular shadow? Draw or describe that position. Go inside and compare your results with others in the class.

II. Now, cut a 4-inch by 5-inch rectangle from the paper and attach a straw to it.

Before you go outside, draw pictures of how you think the shadows will look when you hold the rectangle both parallel and perpendicular to the ground. Go outside and hold the shape parallel and perpendicular to the ground. Draw pictures of your shadows and compare with your guesses. Were you correct?_____

Now, find a position for the rectangle which will produce a square shadow. Draw that position, and do the same for a rectangular shadow!

**TITLE:** **Tab It! (7.07)**

GOAL: Construction of geometric solids is always interesting to students, but sometimes teachers give the students too much help in this task. This activity asks students to put the gluing tabs on flat patterns for geometric figures.

OBJECTIVES: The student will draw patterns (net) for given geometric solids.

The student will fold flat patterns into models of geometric solids.

The student will locate positions for tabs on flat patterns for geometric solids.

MATERIALS: Paper, glue or paste, scissors.

GRADES: 5, 6, 7, 8, General Mathematics.

INSTRUCTIONS FOR TEACHERS: Each student should partake in this activity individually. The materials are few in number and easily found, so all children should have adequate supplies for this purpose.

COMMENTS: Geometric solids such as octohedra, dodecahedra, isocahedra, cones, and spheres make excellent challenging figures for the very able students.

TEACHER COMMENTS TO STUDENTS:

Making patterns to cover geometric solids is fun! Let's try some!

Which of the patterns shown in Figure 7-16 can be folded to form a cube?

That's correct. The third figure will make a cube. The first figure has only four sides and in the second there is no way to cover both the ends.

Simplified, Interesting Ways to Teach Geometry          159

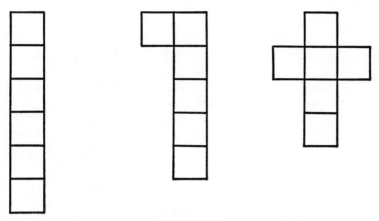

**FIGURE 7-16**

Let's make a pattern like number 3 and cut it out. Fold it to make a cube.

Did you have a problem making it stay together? Where would you put tabs for gluing this cube together? Draw pattern 3 again, and this time put tabs on the pattern.

Compare your results with others. Did you all use the same second pattern?

Now, try to make a pattern with tabs for a tetrahedron, a pyramid, and a rectangular prism. Use as few tabs as possible.

**TITLE:**      **Styrofoam (7.08)**

GOAL: Students should be aware of mathematical reasons for practical situations that exist in life. For example, why does a three-legged table not wobble while a four-legged one might?

OBJECTIVE: Using toothpicks and Styrofoam, the student will discover that three non-colinear points determine a plane.

MATERIALS: Styrofoam, toothpicks, meter stick or ruler.

GRADES: 7, 8, General Mathematics, Geometry.

INSTRUCTIONS FOR TEACHERS: The students need to be encouraged to do this activity on a step-by-step basis. When they begin using three toothpicks, the students should consider the colinear situation.

COMMENTS: This can readily be related to why milk stools have three legs.

TEACHER COMMENTS TO STUDENTS:

Take a Styrofoam rectangle and stick four toothpicks in it. The toothpicks should be perpendicular to the plane of the Styrofoam and near each of the four corners.

By doing this you have created a small table. Measure to be certain that each toothpick has the same length sticking out of the Styrofoam.

Place your small table on a level, flat surface. Do all four toothpicks touch the surface?_____

If all four legs are not touching, how many are?_____ If all four legs are touching, push one farther into the Styrofoam so that one leg is shorter than the other three. Now how many legs touch the surface when you put the small table on it?_____

Push another toothpick into the Styrofoam so that the amount it sticks out is different from either of the other two lengths. When the table is placed on the surface, how many legs are touching? _____Without changing any of the leg lengths, can all four legs be made to touch the surface?_____What is the greatest number of legs touching at any time when they are these lengths?_____

Select one of the remaining two legs that are the same length and push it through to a length different from any of the other legs. Placing the table on the surface and without changing leg lengths, can all four legs be made to touch at once?_____ Regardless of how you place the table, what is the largest number of legs touching (remember you can't change the leg length)?_____

Remove one of the four legs from the table. Can you get the table situated so that only the remaining three legs are touching? _____You might need to move one of the legs so the table will stand on the three legs only.

What geometric word would describe the smooth flat surface you have been placing the table on?_____

How many points are needed to determine the smooth flat surface above?_____ Has that number of points occurred at any other time in this activity?_____

Four points could determine a plane. Describe a situation in which this could happen. _____

Can you verify your response with the Styrofoam and toothpicks? Do four points necessarily determine a plane?_____

Would it be possible that more than four points could determine a plane?_____Can you make a model of your response?

Why do some four-legged tables wobble?_____

**TITLE:** **Geocab (7.09)**

GOAL: Students enjoy activities where they are given arrays of letters in which they are to find words. Often these arrays are of a general nature, but the idea can easily be adapted to productive classroom use.

OBJECTIVE: The student will, given an array of letters, locate and loop geometry vocabulary words.

GRADE: Geometry

MATERIALS: Student Geocab sheet (Figure 7-17)

INSTRUCTIONS FOR TEACHERS: Students need to know that the words are written horizontally, vertically and diagonally. Furthermore, some of the words are in reverse order.

COMMENTS: This activity can be used in a contest manner to see who can get the most words, the largest number of the same word, a specified number of words, the most words within a designated time period, a particular word in the shortest time, etc. Perhaps you would like to make your own Geocab. Determine the size of the array, write in the appropriate words and fill the empty cells randomly. Be sure to check carefully for obscene words.

**TITLE:** **Fold 'em Up I (7.10)**

GOAL: Paper folding can be a useful activity for motivating students. This activity motivates learning and provides a variety of paper-folding activities which can be useful for basic compass-straight-edge constructions.

OBJECTIVE: The student will perform geometric constructions through paper folding.

MATERIALS: Paper or waxed paper, pencil.

GRADES: 7, 8, Geometry, General Mathematics.

INSTRUCTIONS Waxed paper has an advantage over a more standard

*162*  *Simplified, Interesting Ways to Teach Geometry*

| R | E | A | N | I | L | L | O | C | P | B | H | Q | U | O | P | R | I | S | I | O | N | T |
|---|---|---|---|---|---|---|---|---|---|---|---|---|---|---|---|---|---|---|---|---|---|---|
| G | D | E | A | E | X | A | C | O | N | K | Y | A | N | O | I | T | C | E | L | F | E | R |
| M | G | Y | E | H | B | R | A | N | N | C | P | A | C | U | S | D | X | E | C | A | F | B |
| L | E | B | I | L | E | N | P | E | R | H | O | M | B | U | S | A | Q | U | I | L | G | F |
| S | W | I | E | A | V | C | N | H | D | A | T | R | A | A | L | E | E | T | U | T | A | V |
| C | A | C | R | T | O | L | Y | R | A | L | J | H | E | X | O | T | M | E | D | I | A | N |
| S | E | O | N | Y | W | A | S | E | C | A | N | T | I | I | P | A | R | T | L | T | C | L |
| O | T | N | N | O | V | E | T | C | U | R | U | S | V | O | E | F | T | A | F | U | I | L |
| H | Y | D | N | T | E | S | E | O | R | C | S | D | O | M | U | G | L | R | A | D | S | A |
| O | J | I | N | H | A | K | O | N | T | A | E | P | O | S | T | U | L | A | T | E | S | C |
| I | R | T | U | E | R | A | M | C | N | H | G | U | A | B | A | G | M | Y | U | R | B | I |
| S | O | J | N | S | J | M | U | E | E | S | T | D | J | T | N | E | M | G | E | S | I | T |
| T | C | O | S | I | N | E | H | N | U | T | S | B | U | B | G | H | W | U | H | H | H | R |
| C | H | N | A | S | E | S | L | T | R | E | H | S | L | I | E | Z | A | B | Y | E | T | E |
| A | R | A | V | S | I | L | L | R | G | A | Y | M | L | I | N | E | R | I | P | J | T | V |
| R | N | L | O | C | U | S | J | M | I | N | L | P | X | K | T | T | I | H | R | E | I | H |
| R | I | Q | A | I | Z | R | T | C | O | R | O | L | L | A | R | Y | Z | A | R | H | E | C |
| I | N | S | C | R | I | B | E | D | C | G | T | O | D | E | R | F | E | V | F | R | O | E |
| D | V | K | M | R | D | N | X | N | Y | P | E | T | O | I | M | O | N | E | D | O | R | L |
| E | E | E | E | Y | I | E | A | I | S | O | S | C | E | L | E | S | I | C | L | E | E | H |
| A | R | W | T | Q | U | F | H | L | E | I | I | D | O | F | R | U | S | T | R | U | M | N |
| T | S | U | P | O | T | T | C | I | B | A | S | A | H | W | N | Y | W | O | O | N | K | T |
| V | E | R | T | E | X | N | T | T | D | E | S | F | H | A | G | R | U | R | A | L | X | O |

**FIGURE 7-17**

FOR TEACHERS: type of paper. When folded, waxed paper leaves a white line segment on the fold which is quite easily identified when a sharp crease is made. New paper should be used for each activity. Lines can easily be scratched on waxed paper.

COMMENTS: Here is a representative sampling of activities which can be accomplished with paper folding that relate to basic constructions. (Certainly, there are many more activities of this type, as well as more complex and different ones.)

1. "Paper Folding," National Council of Teachers of Mathematics, 1906 Association Drive, Reston, Va., 22091. #301-09152 (90¢).
2. *Patterns and Puzzles in Mathematics*, The Franklin Mathematics Series, Lyons & Carnahan, Educational Division/Meredith Corp., Chicago, Illinois 60616.
3. Cheatham, Ben H., Jr., "A Comparison of Two Methods of Introducing Selected Geometric Concepts to Seventh Grade Students." Unpublished Ph.D. dissertation, University of Florida, 1969.

Paper-folding activities can also be used to teach equivalent fractions.

## TEACHER COMMENTS TO STUDENTS:

Some geometric constructions usually done with a compass and straight-edge can be made by folding paper. You will need some paper and a set of instructions to guide you in these paper-folding activities.

A line perpendicular to a given line can be constructed by (1) folding the paper to make the given line, (2) opening the paper to its original form, and (3) refolding the paper so that part of the given line can be made to coincide with the remaining part of the given line. Then, folding, the desired perpendicular can be constructed.

Pictorially, the procedure would be as illustrated in Figure 7-18.

1. Fold a paper to construct a perpendicular to a given line at a given point on that line.
2. Fold a paper to construct a perpendicular to a given line, the perpendicular containing a point not on the given line.
3. Fold a paper so you get two lines perpendicular to the same line. How will the two lines perpendicular to the same line be

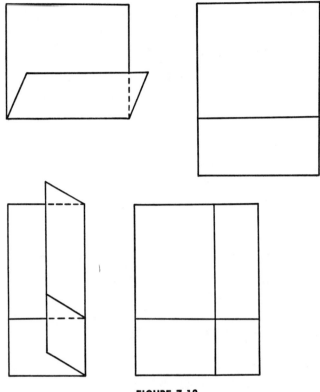

**FIGURE 7-18**

related? Can you make additional fold(s) to verify your answer?
4. Take a sheet of paper which is not rectangular in shape and make folds in that paper so that those folds form a rectangle. Because a rectangle has four sides, you will need to make at least four folds. After you have formed a rectangle, can you make additional folds to check the accuracy of the rectangle?
5. Using procedures similar to those used for the rectangle, fold a square.

**TITLE:** **Fold 'em Up II (7.11)**

GOAL: Paper-folding activities are not necessarily limited to

*Simplified, Interesting Ways to Teach Geometry* 165

|  |  |
|---|---|
|  | basic constructions of lower level mathematics classes. The folding procedures here could be for more capable or advanced students. |
| OBJECTIVE: | The student will form curves, normally derived by cutting a right circular cone with a plane, through paper folding. |
| MATERIALS: | Paper or waxed paper, scissors, compass, ruler. |
| GRADE: | Geometry, Analytic Geometry. |
| INSTRUCTIONS FOR TEACHERS: | The advantage of waxed paper leaving a white line segment when folding to a sharp crease is most apparent here. The more folds the student makes, the better is his approximation of the curve. |
| COMMENTS: | This is only a representative collection of folding activities. It will be advantageous for you to have done these activities before presenting them to the students to ease any clarification of instructions. |

TEACHER COMMENTS TO STUDENTS:

A series of lines can be used to develop curves. The more lines made and the closer together they are, the closer the true curve is approximated. Actually, each of the folds represents a tangent to the curve which is being outlined.

1. A parabola can be formed by placing a point, A, near the center of one edge of a rectangularly shaped piece of paper, the point being three or four centimeters from that edge. Take the paper edge closest to the point A and fold the paper so that the line determined by that edge passes through point A. The place where the edge passes through A determines a point on edge B. After opening the paper make a second fold, different from the first but still having that same edge pass through point A. This fold will determine another point on the edge, this point being different from B. Repeat this procedure of making new folds, taking the edge through point A about 25 times. When completed, the 25 or so folded lines should approximate a parabola. Point A will be the focus of the parabola and the edge folded to pass through A will be the directrix.

Pictorially, the procedure has been as shown in Figure 7-19. Continue making folds until the paper looks like Figure 7-20.

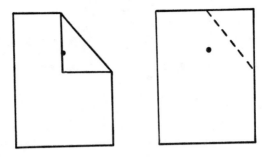

**FIGURE 7-19**

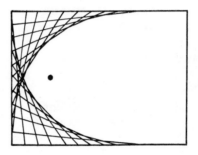

**FIGURE 7-20**

2. Draw a circle approximately one decimeter in diameter on one half of a rectangular sheet of paper. Mark a point, O, about three centimeters outside the circle toward the blank half of the paper. (The line determined by O and the center of the circle must be parallel to the longer edge of the paper.) Fold the paper so that O is on the circle. Repeat this procedure until O has been folded into different positions all the way around the circle.

Upon completing the final fold you should see two branches of a hyperbola. Point O is one focus of the hyperbola; can you find the other focus? (See Figure 7-21.)

3. Draw a circle with a one-decimeter radius and cut the circle out. Mark a point, E, three centimeters from the edge of the circular region. Fold the paper so the circle passes through E. Repeat this procedure several times, being certain that points all the way around

*Simplified, Interesting Ways to Teach Geometry* 167

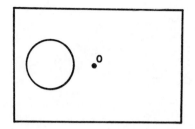 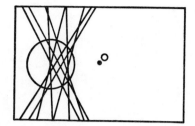

**FIGURE 7-21**

the circle are made to coincide with E. What figure or curve have you outlined with all these folds? Are there any foci for this curve and, if so, can you locate them? If, instead of being three centimeters from the edge, E was located at the center of the circle, and the same procedures were followed, what would the resultant curve be? (See Figure 7-22.)

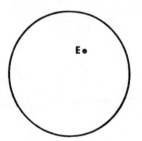 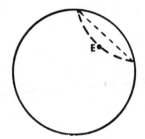 

**FIGURE 7-22**

# 8
# Developing Meaningful Data via Graphs

Graphing is a very useful procedure for illustrating mathematical and statistical information, and most teachers of mathematics include graphing in their mathematics program. However, many of the common problem situations involving graphing in statistics or analytical geometry fail to motivate the students. Thus, the activities included in this chapter have been selected to provide the teacher with meaningful and enjoyable activities involving concepts from graphing.

There are many interesting situations which can be used to generate data for graphing statistical information. The teacher of mathematics who is searching for new ideas will find activities in this chapter based on science, recreation, ecology, medicine, music, and athletics.

| | |
|---|---|
| **TITLE:** | **Milk Jugs (8.01)** |
| GOAL: | Many activities can be generated from common household objects. This helps students realize the concepts involved in rate problems using familiar materials from daily life experiences. |
| OBJECTIVES: | The student will graph results on the Cartesian plane. |
| | The student will predict results based on geographical data. |
| | The student will compute area. |
| MATERIALS: | 5 milk jugs, sand, device to cut holes in jugs. |
| GRADES: | General Mathematics. |

| | |
|---|---|
| INSTRUCTIONS FOR TEACHERS: | CAUTION: The bottoms of the jugs resist cutting and extensive care must be taken while cutting holes. The holes should be cut in approximately the center of the bottom, each jug having just one hole in it. It is helpful to work over a sand box. |
| COMMENTS: | There are several varieties of this activity. For example, does the shape of the hole alter the time required for drainage given that the area remains constant? Or, given a constant shape and area, does the location of the hole in the bottom affect the drainage time?<br>Water could be used rather than sand. It helps to put the lid on the jug and fill it through the hole (a funnel would help), cover the hole with your hand, turn the jug right side up, remove the lid, and then uncover the hole and begin timing drainage. |

TEACHER COMMENTS TO STUDENTS:

You are to cut holes in the bottoms of the milk jugs, putting one hole in each jug. The holes should be: a square 6 mm on a side; a square 1.2 cm. by 1.2 cm.; a triangle of base 1.2 cm. and height 1.2 cm.; a rectangle 1.2 cm. by 6 mm.; and a hole of shape and area of your choice.

Each jug is to be filled with sand and the amount of time required for drainage measured. As the jug drains, the sand will flow out quickly for a time and then the flow will decrease. If you consider the cut-away view of the jug shown in Figure 8-1 you can see why the flow decreases.

**FIGURE 8-1**

You should decide when the sand reaches the level pictured, mark that level on the outside of all the jugs, and stop timing when the sand reaches that level. NOTE: The stopping level should be that of the jug with the smallest area so that all are timed for the same period.

The times for the four known areas should be entered on a graph. These four points should be connected with a smooth curve and that curve should be used to predict the area of the fifth hole since you will know the time required for that fifth jug to drain.

As a check, place a piece of graph paper over the hole and trace the hole, then find the area. The hole can be traced by cutting the jug top off so you can get your hand inside. Or, you could scribble back and forth in a method similar to that used to reproduce one side of a coin. Using the second method requires care to not poke a hole in the paper.

**TITLE:** **How Tall? (8.02)**

GOAL: The physical sciences are often used as sources of activities in mathematics, but the biological sciences can be used also. This activity uses the rapid germination of beans as a source of data for graphing.

OBJECTIVES: The student will compute percents.
The student will make a circle graph.
The student will make histograms.
The student will compute averages.
The student will make a line graph.

MATERIALS: Empty milk cartons, dirt, beans.

GRADES: 5, 6, 7, 8, General Mathematics.

INSTRUCTIONS FOR TEACHERS: This activity is accomplished most successfully with small groups of students responsible for a small set of beans.

Use no fewer than 20 beans. Depending on the quality of the seed and soil the germination rate can result in such a meager production that the graphing is useless. If 50 or 100 beans can be planted, it is better.

The following graphs can be made with the data:

| Bean No. | Germination Day | Day 1 | 2 | 3 | 4 | 5 | 6 | 7 | 8 | 9 | 10 | 11 | 12 | 13 | 14 | 15 | 16 | 17 | 18 | 19 | 20 | 21 |
|---|---|---|---|---|---|---|---|---|---|---|---|---|---|---|---|---|---|---|---|---|---|---|
| 1 | | | | | | | | | | | | | | | | | | | | | | |
| 2 | | | | | | | | | | | | | | | | | | | | | | |
| 3 | | | | | | | | | | | | | | | | | | | | | | |
| 4 | | | | | | | | | | | | | | | | | | | | | | |
| 5 | | | | | | | | | | | | | | | | | | | | | | |
| 6 | | | | | | | | | | | | | | | | | | | | | | |
| 7 | | | | | | | | | | | | | | | | | | | | | | |
| 8 | | | | | | | | | | | | | | | | | | | | | | |
| 9 | | | | | | | | | | | | | | | | | | | | | | |
| 10 | | | | | | | | | | | | | | | | | | | | | | |
| 11 | | | | | | | | | | | | | | | | | | | | | | |
| 12 | | | | | | | | | | | | | | | | | | | | | | |
| 13 | | | | | | | | | | | | | | | | | | | | | | |
| 14 | | | | | | | | | | | | | | | | | | | | | | |
| 15 | | | | | | | | | | | | | | | | | | | | | | |
| 16 | | | | | | | | | | | | | | | | | | | | | | |
| 17 | | | | | | | | | | | | | | | | | | | | | | |
| 18 | | | | | | | | | | | | | | | | | | | | | | |
| 19 | | | | | | | | | | | | | | | | | | | | | | |
| 20 | | | | | | | | | | | | | | | | | | | | | | |

**FIGURE 8-2**

1. A circle graph showing the percent of germination.
2. A histogram showing the number of beans breaking through the soil on given days.
3. Daily histograms showing the number of beans between 0-1", 1-2", 2-3", etc.
4. A line graph showing the average height of the bean plants each day.

COMMENTS: If your school has a garden area, the surviving plants may be transplanted. The garden plants can provide more data for future graphing lessons such as survival rate, height, or production.

*Developing Meaningful Data via Graphs*

TEACHER COMMENTS TO STUDENTS:

Let's plant some beans! First, open the tops of the empty milk cartons and wash them. Fill the clean cartons with rich dirt, and sprinkle with water. Tape a numeral to each carton for identification.

Now, put the beans in a jar of water overnight. We will plant them ½" deep in the dirt tomorrow.

Each day for three weeks water the beans and record data in the chart shown in Figure 8-2.

At the end of three weeks, your teacher will help you organize the information about your beans.

**TITLE:** **Tickets, Please! (8.03)**

GOAL: Many non-reading students have difficulty interpreting graphs even though graphs are mostly nonverbal. This activity gives examples of how histograms are made in which the student becomes actively involved.

OBJECTIVES: The student will construct a histogram.
The student will interpret a histogram.
The student will use ratio to make the scales for a graph.

MATERIALS: Tickets from an athletic contest, chart paper.

GRADES: 8, General Mathematics.

INSTRUCTIONS FOR TEACHERS: The students should engage in a preliminary activity with the whole class before the ticket problem is introduced.

Have each student cut out a picture like the one in Figure 8-3. Make a histogram of how many people

**FIGURE 8-3**

are in families. Have the students paste or pin their house on the chart paper in the column showing the number of people in their families. (See Figure 8-4.)

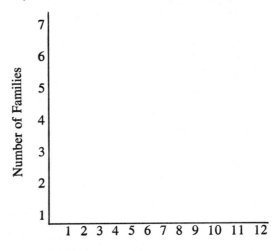

**FIGURE 8-4**

How many in your family?

In the ticket activity the students will need help in establishing a ratio. For example, the group with the largest number of tickets will often want to use a different ratio than the group with the smallest number of tickets. A discussion must take place which will establish a common ratio.

COMMENTS: The students will have difficulty putting the last ticket on the graph because they probably will not use a ratio which requires an even number of tickets for each group. For example, if one ticket represents ten people and the number of tickets is 57, how is the seven represented? You will need to discuss this problem with the entire class.

TEACHER COMMENTS TO STUDENTS:

Let's make a graph showing the attendance at the basketball game.

Use the tickets to make a histogram.

*Developing Meaningful Data via Graphs* 175

Let's get into groups! Make a group for each type of ticket, and count the tickets.

Adult's tickets = _____.
Students's tickets = _____.
Children's tickets = _____.

There are many tickets in each group. Can we make a graph so that every ticket is used?
What can be done to show how many tickets were sold?

---

How many tickets will be shown by one ticket on the graph? Be certain that you have chosen a number that will make the graph fit on your paper!
Glue the tickets on the graph.

## TITLE: Top Ten (8.04)

| | |
|---|---|
| GOAL: | Students in all areas of the country are interested in the most popular records on the market today. This activity exploits this interest to create a graphing lesson. |
| OBJECTIVES: | The student will record data.<br>The student will make a histogram with recorded data. |
| MATERIALS: | Top ten records, tape, and album lists. |
| GRADES: | 6, 7, 8, General Mathematics. |
| INSTRUCTIONS FOR TEACHERS: | Although this is a class activity for the collection of data, all students should make the histograms. |
| COMMENTS: | This activity can be extended in many ways. As a long-term activity, the top ten records, tapes, and/or albums can be recorded over several months, and graphs of the data can be made.<br>Often different radio stations list different songs in the top ten or list the same songs in a different order. Comparing these charts will provide graphing data.<br>What is most popular in the class—records, albums or tapes? |

TEACHER COMMENTS TO STUDENTS:

What tape is most popular in your class?
What record is most popular in your class?
What album is most popular?

Make a chart to show the results of a class poll on these questions. List the top ten for each category—tapes, records, albums—and the number of votes each one gets within its class.

Make a histogram showing your results.

Now, let's set up a new system of voting. Use the results of your first poll plus one more rule. Add two first-place votes for each record and album for each student who owns the record. Did the order of popularity of the records and albums change with this new system of voting? Which system of voting do you think is most fair? Why?

Make a histogram showing the new results!

## TITLE: Pulse Rate (8.05)

GOAL: Students are fascinated with facts about their persons, and this activity which involves taking pulses offers many opportunities to study mathematical relationships.

OBJECTIVES: The student will record data collected in a group activity.
The student will graph the data collected in a group activity.
The student will state generalizations formed from observations of the collected data.

MATERIALS: Sweep second hand clock or watch.

GRADES: 5, 6, 7, 8, General Mathematics.

INSTRUCTIONS FOR TEACHERS: Separate the class into groups of four to six students each and include members of both sexes in each group. By including both boys and girls in the group, you may generate some interesting relationships in the data such as: "John made all the girls' pulses go up!"

COMMENTS: Further patterns in the data can be generated by having the students exercise mildly and repeat this activity. The increase in pulse rate due to exercise will be interesting to the students.

Four graphs could be developed with this activity, each of which would be a broken line graph on a separate transparency. The data determining each graph would be boys' pulses taken by boys, girls' pulses taken by girls, boys' pulses taken by girls, and girls' pulses taken by boys. The four transparencies could then be superimposed on an overhead and differences and similarities discussed.

TEACHER COMMENTS TO STUDENTS:

Using a sweep second hand on a wall clock or a wrist watch, take the pulse of some of your classmates.

Remember, count one heartbeat for each throb, and do this for one minute.

Record your own results, showing the sex of the persons taking your pulse and the pulse rate they get for you.

Now, each person in the group will make a vertical bar graph of his pulse taken by the other members of the group. Put the pulse rate values on the vertical scale and the different pulse takers across the bottom.

Compare your graph to the graphs of others in your group. What generalizations can you make?

**TITLE:** **Car Pool (8.06)**

GOAL: Newspaper, radio and television commentaries now encourage formation of car pools. Industries sometimes encourage employee participation in express bus routes from a number of population centers to their plant. Conservationists and anti-pollutionists discuss the environmental damage caused by combustion emissions. Are people in general changing their habits as a result of these warnings? This activity will allow students to investigate the impact of these warnings.

OBJECTIVES: The student will gather traffic data and make a frequency distribution.

|               |                                                   |
|---------------|---------------------------------------------------|
|               | The student will gather traffic data and make a histogram. |
| GRADES:       | 7, 8, General Mathematics, Senior Mathematics.    |
| MATERIALS:    | Clipboard, pencil and paper.                      |
| INSTRUCTIONS FOR TEACHERS: | To provide background, observe traffic as you go to or from school for a few days. This will provide you with a rough estimate of how many vehicles in your area are occupied by a single individual. |
| COMMENTS:     | Perhaps a survey could be taken within the class or school by you or the students to determine how many parents drive to work alone, how many are in car pools or are a steady driver for a group, and how many could form car pools with co-workers. This activity could be easily extended to a traffic flow survey similar to those done in departments of transportation. Conceivably, a transportation representative could discuss procedures to be followed with the class. |
|               | Local efforts to reduce the number of vehicles on the road should be pursued. Checks at the same points at times other than rush hours might provide some interesting contrasts or similarities. |

### TEACHER COMMENTS TO STUDENTS:

Recently many requests have been made through the media to cut down the number of vehicles on the highways. In this activity you will determine two things: how many vehicles are on a given road during a certain time period, and how many of those vehicles are occupied by a single individual.

You are to select a road, preferably one which is a main traffic artery for industry or business and preferably during the times when people are going to or from work. You are to count the cars going in one or both directions and note how many of them have only one person in them.

A frequency distribution is a summary of information. In this case it would be a chart like the one shown in Figure 8-5, only the spaces would be filled in. Information pertaining to where the data was gathered, date and time of day should be listed with the frequency distribution to aid in its interpretation.

## Developing Meaningful Data via Graphs

| Number of people in car | Frequency | Ratio | Percent |
|---|---|---|---|
| DATE_____ TIME_____ PLACE_____ | | | |
| 1 | | | |
| 2 | | | |
| 3 | | | |
| 4 | | | |
| 5 | | | |
| 6 | | | |

**FIGURE 8-5**

A histogram is like a bar graph. Using the data from the frequency distribution, make a histogram using the outline presented in Figure 8-6. Note that each bar is the same width.

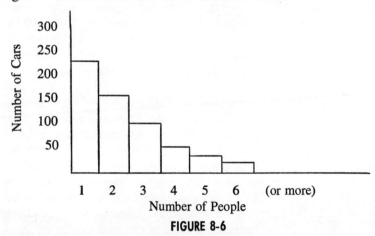

**FIGURE 8-6**

**TITLE:** **Toboggan Run (8.07)**

**GOAL:** The student will measure the speed of a toboggan (sled) carrying people of different masses.

**OBJECTIVES:** The student will generate data to be used in graphing the relationship of the speed of a toboggan and its weight.

The student will measure mass, distance and time.

| | |
|---|---|
| GRADES: | 7, 8, 9, 10, 11, 12. |
| MATERIALS: | Tape measure, stopwatch, scales, paper, pencil, sled or toboggan, a hill with at least a 50-yard toboggan run. |
| INSTRUCTIONS FOR TEACHERS: | 1. You may wish to refer to the math lab "Speed Trap" in *School Science and Mathematics*, February, 1974, to become more familiar with the intricacies of measuring average speed (Activity 2.03).<br>2. The toboggan run should be used several times to establish a track which will help to control for the differences in speed caused by the friction in fresh snow.<br>3. Groups of students will be best for this laboratory. This allows for two students to serve as timers, while four may take turns riding the sled individually or in groups. |
| COMMENTS: | This experiment could also be done with a skate board or wagon. For some classes, concluding remarks about rules of soapbox derbies may increase the relevancy of the work. For example, there is a weight limit placed on the total weight of the car and the child. Why?<br>If it is not possible to get outside, this experiment can be conducted in your classroom. Stretch a string between two points (one higher than the other). A pulley should be on the string and it must be possible to attach a weight to the pulley. The descent of the pulley and weights down the string will simulate the toboggan with different weights. Perhaps some students could investigate the results of changing the angle of elevation or rope tension. |

TEACHER COMMENTS TO STUDENTS:

Given the same toboggan and the same toboggan run, will a 200-pound football player or a 100-pound girl riding on the sled make it go faster?

To answer this question, go to the toboggan run!

First, set up a way to calculate the speed of the toboggan. Put one person at the beginning of the run and one at the end. The first person will signal the person at the finish line when the toboggan begins moving. The second person starts the stopwatch at this time and times the sled as it crosses the finish line.

Time several individuals or groups of students on the toboggan and record the weight, time and speed for each in a chart.

In your calculations of the speed, you will be computing the average speed of the toboggan. The time and distance measured are used in the formula for average speed, $d = rt$. Be sure to convert the distance to miles and the time to hours if you wish to express your answers in miles per hour.

Now, graph the results of your work! From the information on the graph, try to make a generalization about the relationship between the speed of a toboggan and the weight it carries.

**TITLE:** **The Right Slant (8.08)**

GOAL: Slope can be explained by lecture demonstration or by involving the student in a programmed type of situation. This activity will provide a broad background of information from which the student can develop a concept of slope.

OBJECTIVE: The student will recall the meaning of slope.

GRADES: Algebra 1, Pre-Algebra, General Mathematics.

MATERIALS: Student hand-out, graph paper, straight-edge.

INSTRUCTIONS FOR TEACHERS: In this activity the student is given characteristics of the slope as an aid for the development of a definition. Examples include both positive and negative slopes, but there is a need for more of each type. Comment is made about the slope being the same regardless of the points used on the line.

COMMENTS: This activity can be used as a variation of the plotting of points to derive a picture. Segments of lines can be plotted giving the desired pictures.

Further consideration can also be given to the described situation. Questions like, "What if you go left from the line, then up?" could be treated. The discussion in this activity has the triangle giving the x- and y-value being formed below the line. Understanding could be enhanced by providing a different orientation, having the triangle above the line.

TEACHER COMMENTS TO STUDENTS:

1. On graph paper draw x- and y-axes.
2. Connect the points (5,5) and (0,0) with a line.
3. How many units up from the x-axis is the point (5,5)?_____
4. How many units over from the y-axis is the point (5,5)?_____
5. What is the ratio of y-units to x-units?_____
6. Draw another set of x- and y- axes.
7. Connect the points (4,8) and (0,0) with a line.
8. How far in y-units is the point (4,8) from the x-axis?_____
9. How far in x-units is the point (4,8) from the y-axis? _____
10. What is the ratio of y-units to x-units?
11. What can we say about the second line?
12. We call this characteristic of the line its *slope* and the ratio of y-axis to x-units is the numerical value of the slope.
13. Draw another set of x- and y-axes.
14. Connect the points (-4,8) and (0,0) with a line.
15. How far in x-units is the point (-4,8) from the y-axis?_____
16. Is this a positive or negative directional value of x? _____
17. How far in y-units is the point (-4,8) from the x-axis? _____
18. Is this a positive or negative directional value of y? _____
19. What is the ratio of y-units to x-units? _____
20. Why? (referring to # 19)_____
21. What can we say about a line which has a negative slope?
22. What can we say about a line which has a positive slope?
23. On the first graph done in this activity, suppose that you start at the point (2,2) and go to the right two units. Then go up until you meet the line. How many y-units would you go up? _____
24. What is the ratio of y-units to x-units?
25. How does this ratio compare with the slope?

26. How does this ratio compare with slope's initial ratio found from (0,0) to (5,5)?
27. This indicates that any pair of points on a line can be used to get the slope by starting at the lower point and moving parallel to the x-axis until directly beneath the second point, getting the x-value for the slope ratio. The y-value for the slope ratio is the distance between the point below the second point and the second point itself. The ratio of x-value to the y-value is the _____ of the line.
28. Go to the right of a certain number of x-units from a point on the line and then *down* to meet the line. What can we say about the slope?
29. The distance to the right or left from the line is called the *run*.
30. The distance up or down to meet the line is called the *rise*.
31. Going up to meet the line, is the rise positive or negative? _____ Going down? _____
32. Can you state a relationship which defines the slope in terms of the rise and the run?

## TITLE: Noise Pollution (8.09)

**GOAL:** Noise is constantly with most of us and yet much of it goes uninterpreted. This activity will be concerned with the interpretation of noise generated by fans at an athletic contest.

**OBJECTIVES:** The student will graph data gathered at an athletic contest.

The student will attempt to describe the game action by listening to tapes of the crowd reactions.

**MATERIALS:** Three tape recorders, decibel measurer, stopwatch.

**GRADES:** 9, General Mathematics.

**INSTRUCTIONS FOR TEACHERS:** A tape recorder should be located within each of the two cheering sections and left on throughout each half or quarter. The third recorder should be used by a person who gives a play-by-play description of the game. The scoreboard clock time should be read onto all tapes every two minutes for synchronization purposes.

COMMENTS: The sports selected for use should be those that generate noise in bursts, but these bursts should be relatively frequent. Basketball is probably the best choice because of the enclosed area, rapid changes in action, frequent controversial officiating calls, and the closeness of the fans to the action.
This activity could also be performed on busy streets to measure traffic noise.

TEACHER COMMENTS TO STUDENTS:

Is the crowd noise at a basketball game helpful in describing the action? From the group who will be doing this activity, at least one person must sit in each of the two student cheering sections. That student or group of students will take a tape recorder and record the noise within that section throughout the game.

When the game starts, begin recording and leave the recorder running except for time-outs. Since other tape recorders will be involved, to aid in synchronizing the tapes read the scoreboard time at even minutes onto the tape.

A third person or group of persons will give a detailed play-by-play description of the action as the game progresses, also reading the even minutes onto the tape from the scoreboard. This tape will be used to determine if the student cheering noise can help describe the action.

When analyzing the tapes, a stopwatch or a watch with a sweep second hand will be needed. Every 30 seconds the intensity of the sound, as read from the decibel measurer, is to be recorded for each tape.

Taking a sheet of graph paper, place the axes system to the left and bottom, leaving only enough room for the labels and the scales. One axis should indicate the loudness and the other the time intervals.

First plot the data from the home team tape and join the points with a smooth curve. Then plot the visitor's tape data on the same piece of paper, joining this set of points with another smooth curve, preferably of a different color. Using the play-by-play tape, indicate which team (home or visitor) was in control of, or providing the most significant action for, each 30-second interval. It may be necessary to play the tape back to determine which team should be listed at some intervals.

Are the graphs high and low at the same time? _____
When one curve is high is the other one low most of the time?
_____

Does the indicated control of action agree with the high points of the curves (assuming that the high points indicate more noise generated)?

Is there any relation between the curve with the most peaks (high points) and the winner of the game? In other words, did the team with the most cheering peaks win? _____

**TITLE:** **Rain (8.10)**

GOAL: Children frequently ask how one can tell the amount of rain which has fallen in a period of time. Weather forecasters list the rainfall per day to hundredths of an inch. How do they get the values they use?

OBJECTIVES: After acquiring or constructing a rain gauge, the student will measure the rainfall at his home on a daily basis for a month.

The student will plot the information gathered with his rain gauge on a coordinate grid.

The student will compute the average rainfall at his home over a given time period.

Using the rainfall information and that of others, the student will compare rainfall tendencies for different geographic regions.

MATERIALS: If a gauge is acquired: graduated cylinder or beaker.

If a gauge is constructed: tuna can, plastic funnel, (inside diameter of funnel equal to outside diameter of tuna can), thin, tall jar or graduated cylinder, stand to hold the rain gauge, and rain!

COMMENTS: This activity is quite versatile since it could involve application of statistics, graphing, measurement, computation and construction. The situation can also be applied in snow country by knowing the conversion rate for snow to water. Students comparing rainfall data with that of classmates might find rather large variations. Perhaps the students or teacher could convince friends in another town to duplicate the activity, thus providing more data for comparison. The students could compare the open rainfall to that under a tree.

TEACHER COMMENTS TO STUDENTS:

Have you ever had a friend tell you it rained hard at his house, and it barely sprinkled at yours? Have you seen rain falling a block away from you and yet you remain dry because it isn't raining where you are? How common are such situations? This activity will provide useful information for answering questions such as these.

You will make a rain gauge that should look like the one in Figure 8-7.

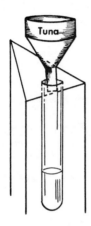

**FIGURE 8-7**

The tuna can (or whatever) should fit inside the funnel and serve as a guard to keep water from splashing out. Since the funnel and can are placed inside the clear, thin jar which has a small base, the stand is necessary to avoid the tipping of the assembly.

Once assembled, the rain gauge should be placed in the open so that it will not collect rain that drips from a tree, roof, etc. The gauge should be emptied and the amount of water recorded after designated periods of time (usually every day) to insure that the basis for comparison is the same. Since the tuna can is larger than the jar, the inside diameters of the two must be measured. Establish a ratio of the area of the bottom of the jar and the bottom of the tuna can. This ratio is necessary because rainfall is reported as a depth on the surface, assuming none runs off or soaks in. Since the collector (tuna can) is larger than the holder (jar), allowances must be made to compensate for the amount gathered.

For example, suppose the tuna can had a four-inch inside

# Developing Meaningful Data via Graphs

diameter (ID) and the jar had a two-inch ID. Recall that the comparison is between areas so that the base area of the tuna can is $\pi r^2$ or $\pi(2)^2$ and the base area of the jar is $\pi r^2$ or $\pi(1)^2$. The ratio is $\frac{\pi(1)^2}{\pi(2)^2} = 1/4$. That is, whatever the depth is in the jar, it must be multiplied by 1/4 to determine the amount of rainfall on that area.

Assume you want to determine the daily rainfall for a week. Empty the gauge each morning and record the results on a graph.

Recall that the mean, or average, can be computed by adding the values and dividing by the number of addends. In this case, you would add the rainfall for each day to get the total for the week and then divide by seven to get the average daily rainfall. What was the average daily rainfall at your place? How does your average daily rainfall compare with that of other people in the class? If the averages are similar, were the daily values similar? Can you find a person who had some daily values close to yours? Why do you suppose this difference exists? Try the activity for a longer period of time and see if the difference becomes smaller.

# 9
# Activities That Teach Number Theory to All Students

Topics in number theory are often reserved for the more able students as sources of enrichment lessons. However, the activities provided in this chapter will allow teachers of mathematics to present number theory topics to students of all ability levels. For students of lower ability it is often necessary to present topics on the concrete level; thus, each of the activities allows the students to manipulate materials to arrive at a generalization about number patterns.

Experienced teachers of mathematics are well aware of the fact that the patterns in mathematics are often intrinsically motivating to students. The beauty of mathematics inherent in patterns such as the Fibonacci sequences, or the power of patterns indicated in simple geometric sequences, are fascinating to all students. These activities will provide the teacher with the means of exposing all students to the joy of discovering patterns in number theory.

Not only do activities in number theory provide motivation and enrichment, but they also include a great amount of computational practice. To discover the patterns generated by triangular numbers, for example, students must add, subtract, multiply and divide whole numbers. Thus, students of low ability will be practicing basic skills while enjoying mathematics activities involving number theory.

## TITLE: Rat Maze (9.01)

**GOAL:** This activity will introduce the student to the Fibonacci sequence through an experimental activity.

**OBJECTIVE:** The student will generate the Fibonacci sequence by recording data derived from a "rat" running a maze.

**MATERIALS:** A stuffed toy mouse, a set of 10-12 hexagonal containers (egg carton cups).

**GRADES:** 8, General Mathematics, Algebra 1.

**INSTRUCTIONS FOR TEACHERS:** This activity has been written for 8th graders and General Mathematics students. Students of high ability may be able to do this activity by using drawings of the maze rather than the actual maze.

**COMMENTS:** The Fibonacci sequence can be generated from practical situations. Fibonacci himself developed this sequence: 1, 1, 2, 3, 5, 8, 13, 21, 34, 55, 89, ....., when he considered a specialized breeding problem involving rabbits.

Assume that rabbits start to bear young two months after their own birth, producing only male-female pairs. If one male-female pair is placed in a pen, none of the rabbits die, and they produce an average of one male-female pair each month of their mature lives, how many rabbits will be produced after one year?

Nature provides another source of this interesting sequence in the spiral arrangement of seeds in certain varieties of sunflower and pine cones.

### TEACHER COMMENTS TO STUDENTS:

Henry the rat wishes to go through the maze in Figure 9-1 so that he can eat the cheese.

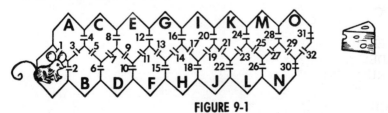

**FIGURE 9-1**

*Activities That Teach Number Theory to All Students*        *191*

Although Henry does not have to go through the maze in alphabetic order, he must always proceed alphabetically toward O. That is, he can go directly from his starting position to B and from there to either C or D, but he cannot go from B to A. Similarly, if he is at K, he can go to M or L, but he may not go to I or J since both I and J come before K in the alphabet.

So, Henry can go through door 1 only to get to room A. If he went through doors 2 and 3 to get to A, he would be going backwards in the alphabet, which is illegal.

How many ways can he get to room B? _____

[Path (door 1, door 3) and path (door 2) will lead to B legally. Thus there are two paths to room B.]

How many ways to room C? _____

[(1, 4), (1, 3, 5), (2, 5)]

Fill in the chart shown in Figure 9-2.

| To Room | Paths | No. of Paths |
|---|---|---|
| A | (1) | |
| B | (1, 3) (2) | |
| C | (1, 4) (1, 3, 5) (2, 5) | |
| D | | |
| E | | |
| F | | |
| G | | |

Can you find a pattern? If so, try it on H through O!

**FIGURE 9-2**

**TITLE:**         **Sum Pages (9.02)**

GOAL:             The student needs to be led to investigate the mathematical possibilities of situations in which he would be prompted to say, "Well, how about that!" This activity pursues one such event!

OBJECTIVE:        The student will establish a pattern from data gathered from a paper-folding activity.

MATERIALS:        Several sheets of paper.

GRADES:           7, 8, General Mathematics.

INSTRUCTIONS FOR TEACHERS:   The student should be encouraged to do several pages of this to be certain he has established the correct pattern.

Be certain the student understands that he will be renumbering pages.

COMMENTS: This activity can be offered independently or with related number pattern situations.

TEACHER COMMENTS TO STUDENTS:

Take a sheet of standard notebook paper and fold it into two congruent pieces, the fold being parallel to one of the edges. Place the paper in front of you on your desk as though it were a book you were about to open and read. The folded edge should be on your left. Label that front page number one and then open the paper out flat, having the *one* facing the table. Label the left page *two* and the right page *three*. Then close the "book" and label the last page *four*. (See Figure 9-3.)

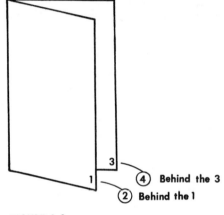

**FIGURE 9-3**

Lay the paper unfolded on your desk. What is the sum of the page numbers showing?_____What is the sum of the page numbers not showing?_____Record these values in a table.

Scribble out the numeral on each page and take a second sheet of paper folded the same way as the first. Put the two sheets together (one inside the other) making an eight-page book, and consecutively number the pages one through eight. Take the book apart after you have numbered each page and find the sum of the numerals on each side of a sheet. The sums for each side of each sheet are_____, _____, _____and_____. Are the sums all the same?

_____They should be, so if they are not, check your work to find the error. Record the results. (When recording the results, list the number of sheets, the number of pages, and the sum of the page numerals on one side of a sheet.)

Repeat the activity with three sheets of paper and again record the results. Each additional sheet adds how many pages to the book? _____ If you knew the number of sheets to be used in a book, how many pages would the book have? _____

How is the sum of the page numbers on a sheet related to the number of pages?_____

Assuming you know the number of sheets to be used in a book, write a formula for the sum of the page numbers on one side of a sheet. _____

| | |
|---|---|
| **TITLE:** | **Mod Art (9.03)** |
| GOAL: | Students are often exposed to the concept of modular number systems for the purpose of exploring the properties of numbers in a finite number system. However, the students are rarely motivated to continue the study of these systems and their properties. This activity will provide the student with leisure time enjoyment of modular number systems and an opportunity to reinforce the properties of operations involving modular numbers. |
| OBJECTIVES: | The student will make an addition table for Mod 6.<br>The student will make a multiplication table for Mod 6.<br>The student will design an artistic representation for the addition and multiplication tables in Mod 6. |
| MATERIALS: | 144 cubes, paint, small paint brushes. |
| GRADES: | 5, 6, 7, 8, General Mathematics. |
| INSTRUCTIONS FOR TEACHERS: | The cubes must be prepared before beginning this activity. Only one pattern for painting the faces of the cubes is given on the student sheets, but there are many other possibilities, such as those shown in Figure 9-4. However, only one pattern should be used on any one set of 144 cubes.<br>This activity is designed for an activity center or a laboratory, and it can be done independently. |

**FIGURE 9-4**

COMMENTS: This activity can take the form of an artistic endeavor. For a reference on paintings exhibiting the structure of mathematics, consult *Interpretations of Painting Exhibiting Mathematical Structure* by Andria Troutman and Sonia Forseth.[6] This teacher's guide to "Math/Art Posters" is published by Creative Publications, Inc.

TEACHER COMMENTS TO STUDENTS:

Complete the table given in Figure 9-5 for Modulo 6.

| + | 0 | 1 | 2 | 3 | 4 | 5 |
|---|---|---|---|---|---|---|
| 0 | 0 | 1 | 2 | 3 | 4 | 5 |
| 1 |   |   |   |   |   |   |
| 2 |   | 3 |   |   |   |   |
| 3 |   |   |   |   | 1 |   |
| 4 |   |   | 0 |   |   | 3 |
| 5 |   |   |   |   |   |   |

**FIGURE 9-5**

Now select six cubes from the set of 144 cubes. Choose one pattern on the faces of each cube to represent each numeral in Mod 6. For example, see Figure 9-6.

Now, using these representations of the modular numbers of Modulo 6, make the rows and columns of the sums in the addition table. How many cubes will you need?_____(Please do not disturb the square representing the addition table until after the next activity.)

---

[6]Forseth, Sonia and Andria Price Troutman, "Using Mathematical Structures to Generate Artistic Designs." *The Mathematics Teacher*, Reston, Virginia: The National Council of Teachers of Mathematics, Inc., May, 1974, pp. 393-398.

# Activities That Teach Number Theory to All Students

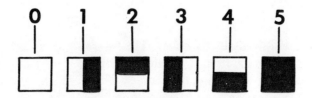

**FIGURE 9-6**

Complete the table in Figure 9-7 for Modulo 6.

| x | 0 | 1 | 2 | 3 | 4 | 5 |
|---|---|---|---|---|---|---|
| 0 | 0 |   | 0 |   |   |   |
| 1 | 0 |   |   |   |   | 5 |
| 2 | 0 |   | 4 | 0 |   |   |
| 3 | 0 | 3 |   |   | 0 |   |
| 4 | 0 |   |   |   | 4 |   |
| 5 | 0 |   | 4 | 3 |   |   |

**FIGURE 9-7**

Using the representations for 0, 1, 2, 3, 4, and 5 that you used in the addition table, make the multiplication table for Modulo 6 with 36 more cubes.

Now, let's check both of these charts for some properties.

1. How can we determine if Mod 6 is commutative for addition and multiplication?_____
   Is addition in Mod 6 commutative?_____ Is multiplication in Mod 6 commutative?_____
2. If there is an identity element for addition in Mod 6, what is it?_____
3. If there is an identity element for multiplication in Mod 6, what is it?_____
4. Show, using patterns, that addition is associative. (See Figure 9-8.)

☐ + (▆ + ▉) $\stackrel{?}{=}$ (☐ + ▆) + ▉

**FIGURE 9-8**

### An Art Activity

You will need all 144 cubes. The following directions use the patterns given in the first part of this activity. Use the addition table as in Figure 9-5, and reflect the addition table across the y axis. This will put the mirror image of the addition table in the first quadrant. Use 36 cubes to make this mirror image. (See Figure 9-9.)

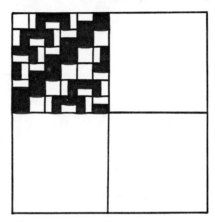

**FIGURE 9-9**

Now reflect the mirror image in the first quadrant over the x-axis into the fourth quadrant. Make this 36-cube pattern.

Now, complete the picture with 36 more cubes by reflecting the original addition table into the third quadrant.

You may want to change the assignment of the patterns to the number in Modulo 6 and make your own work of art. Try making art work using the multiplication table.

### TITLE: Let's Get 'em (9.04)

| | |
|---|---|
| GOAL: | Many students are familiar with members of their athletics teams shaking hands with each other before an athletic contest. This activity utilizes this phenomenon as a motivating device to engage in an activity involving ordered pairs and series. |
| OBJECTIVE: | The student will form ordered pairs to solve a problem involving series. |
| MATERIALS: | Groups of students. |
| GRADES: | 8, Algebra I, Algebra II. |

## Activities That Teach Number Theory to All Students

| | |
|---|---|
| INSTRUCTIONS FOR TEACHERS: | This is an activity which can be used to supplement a lesson on patterns, sequences, or series for junior high school students or general mathematics students. The students may merely form ordered pairs and count to obtain the solution to the problems in this activity. However, Algebra II students may be challenged to determine a formula for solving this problem. |
| COMMENTS: | This is an adaption of the arithmetic series $1 + 2 + 3 + \ldots + (n + 1) + n$ where n = the number of players on a team. When n becomes so large that addition becomes tedious, Gauss has provided a way to generate a formula for this. |

When n is even, we find that by adding the first and last terms, the second and the next-to-last term, etc., the sum is the same for all pairs.

$$1 + n = n + 1$$
$$2 + (n - 1) = n + 1$$
$$3 + (n - 2) = n + 1$$

Since there will be $\frac{n}{2}$ of these pairs, $\frac{(n + 1) n}{2} = \Sigma 1 + 2 + 3 + \ldots + (n - 2) + (n - 1) + n$.

If n is odd, there are $\frac{(n - 1)}{2}$ pairs whose sums are $n + 1$, and one number left over. Since this number left over is the "middle" number, it is $\frac{(n + 1)}{2}$. So $\Sigma 1 + 2 + 3 + \ldots (n + 2) + (n - 1) + n = (\frac{n - 1}{2} \bullet n + 1) + \frac{n + 1}{2}$.

Of course, there are varieties of these formulas, but let the students find them.

### TEACHER COMMENTS TO STUDENTS:

Have you noticed the basketball team slapping hands or giving soul handshakes to their teammates as they are introduced before the game? Let's look at this activity more closely!

1. Bob is introduced and runs onto the court.
2. John is introduced and runs onto the court, and shakes hands with Bob.
3. Doug is introduced and runs onto the court, and shakes hands with Bob and John.

4. Harvey is introduced and runs onto the court and shakes hands with Bob, John and Doug.
5. Slim is introduced and runs onto the court and shakes hands with Bob, John, Doug and Harvey.

How many handshakes are involved?_____ List all the pairs for handshaking.

(John, Bob), (Doug, Bob), (  ,  ), (  ,  ), (  ,  ), (  ,  ), (  ,  ), (  ,  ).

How many were there?_____

1. Let's try this together! Put 11 students in front of the room to be the football team. Have the "team" shake hands as the basketball team did, and keep a record of the hand shaking.
2. How many hand shakes were involved in this example?
3. Try to find a short cut to solve this problem.
4. If you could not think of a short cut, look at the basketball team example again!
   - How many hands did John shake?_____
   - How many hands did Doug shake?_____
   - How many hands did Harvey shake?_____
   - How many hands did Slim shake?_____
   - How many shakes were involved all together?_____
   - Write your short cut here!_____
5. Try your short cut on the football example!
6. Did it work? If not, look for a new pattern for the basketball team example and try again!

**TITLE:** **Prime Numbers (9.05)**

GOAL: Arrays often provide students a new way to look at familiar concepts. Prime numbers will be discussed in this activity using arrays.

OBJECTIVES: The student will make square and rectangular arrays with given numbers of concrete objects.

|              |                                                          |
|--------------|----------------------------------------------------------|
|              | The student will identify prime numbers.                 |
|              | The student will write a definition of prime numbers.    |
| MATERIALS:   | Beans, bingo chips, poker chips.                         |
| GRADES:      | 6, 7, 8, General Mathematics.                            |
| INSTRUCTIONS FOR TEACHERS: | In this activity each student should manipulate the concrete objects and write a number sentence describing each array. The discussions concerning identification of prime numbers and the definition of prime numbers can be conducted with the whole class. |
| COMMENTS:    | This activity provides much practice of the basic multiplication facts. Be sure that all students take part in this phase of the activity. |
|              | This activity may be extended to include discussion of factors of numbers by listing the factors shown in the arrays for each number. |

TEACHER COMMENTS TO STUDENTS:

Numbers may be shown in array form as squares and rectangles. For example, nine beans can be used to form the arrays shown in Figure 9-10.

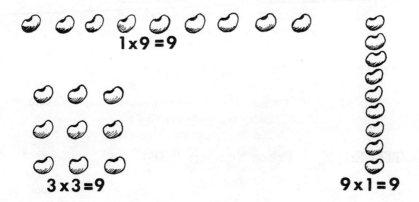

**FIGURE 9-10**

Fill in the chart in Figure 9-11, showing the number of *different* rectangular or square arrays for each of the numbers.

| Number | 1 | 2 | 3 | 4 | 5 | 6 | 7 | 8 | 9 | 10 | 11 | 12 | 13 | 14 | 15 | 16 |
|---|---|---|---|---|---|---|---|---|---|---|---|---|---|---|---|---|
| Number of Arrays | 1 | 2 | 2 | 3 | | | | | 3 | | | | | | | |

| Number | 17 | 18 | 19 | 20 | 21 | 22 | 23 | 24 | 25 | 26 | 27 | 28 | 29 | 30 |
|---|---|---|---|---|---|---|---|---|---|---|---|---|---|---|
| Number of Arrays | | | | | | | | | | | | | | |

**FIGURE 9-11**

How many different arrays do square numbers have? _____ Is there a pattern? _____

There seem to be some other numbers that have only two different arrays. Make a list of these numbers.

_____ , _____ , _____ , _____ , _____ .

What are the next three numbers which have only two different arrays after 29? _____

These numbers are called prime numbers. Write a definition for prime numbers using arrays in your definition.

Numbers having more than two arrays are called composite numbers. Is *one* a composite number? Is *one* a prime number?

**TITLE:**     **Square Numbers (9.06)**

| | |
|---|---|
| GOAL: | Operations on natural numbers often produce very interesting patterns. This activity allows the student to explore the properties of square numbers. |
| OBJECTIVES: | The student will make square arrays. |
| | The student will identify square numbers. |
| | The student will formulate a generalization which produces any square number. |
| MATERIALS: | Beans, bingo chips, or poker chips. |
| GRADES: | 5, 6, 7, 8, General Mathematics, Algebra I. |
| INSTRUCTIONS FOR TEACHERS: | In this activity each student should manipulate the bingo chips. However, discussions concerning the identification of square numbers should be conducted in small groups or as a whole class. |

## Activities That Teach Number Theory to All Students 201

COMMENTS: This activity contains a great deal of computational practice. Thus, be certain that all students manipulate the concrete objects and do all the computational procedures.

TEACHER COMMENTS TO STUDENTS:

Take four bingo chips from your pile! Make a square with the four chips.

Now, take nine bingo chips from your pile and make another square. Did the square look like this? (See Figure 9-12.)

**FIGURE 9-12**

Try to make a square with 16 bingo chips! How many rows will there be?_____ How many columns?_____ Now, try to make a square with 20 bingo chips!

Use the bingo chips and your answers above to fill in the chart shown in Figure 9-13.

Square Numbers

| Number | 2 | 3 | 4 | 5 | 6 | 7 | 8 | 9 | 10 | 11 | 12 | 13 |
|---|---|---|---|---|---|---|---|---|---|---|---|---|
| Square | No | No | Yes | No | | | | | | | | |

| Number | 14 | 15 | 16 | 17 | 18 | 19 | 20 | 25 | 30 | 36 | 40 | 49 |
|---|---|---|---|---|---|---|---|---|---|---|---|---|
| Square | | | | | | | | | | | | |

**FIGURE 9-13**

What will be the next square number after 49?_____
What patterns do you see in the chart?_____

Look at this pattern!

$1 = 1$
$1 + 3 = 4$
$1 + 3 + 5 = 9$
$1 + 3 + 5 + 7 = \square$
$1 + 3 + 5 + 7 + 9 = \square$
$\overline{\phantom{xxxxxxxxx}} = \square$
$\overline{\phantom{xxxxxxxxx}} = \square$

Write a sentence which describes this pattern.

Write a formula which will allow you to compute any desired number of consecutive odd natural numbers.

## TITLE: Pentagonal Numbers (9.07)

GOAL: Operations on natural numbers often produce patterns which are interesting. This activity allows the student to explore the properties of pentagonal numbers and the pattern they generate.

OBJECTIVES: The student will make pentagonal arrays.
The student will identify pentagonal numbers.
The student will formulate a generalization which could be used to produce any pentagonal number.

MATERIALS: Beans, bingo chips, poker chips.

GRADES: 5, 6, 7, 8, General Mathematics, Algebra I.

INSTRUCTIONS FOR TEACHERS: In this activity each student should be given poker chips. It is very important that each person have the opportunity to manipulate the concrete objects. However, the discussions which lead to identification of pentagonal numbers and a general formula for pentagonal numbers could be conducted in small groups or as a whole class.

*Activities That Teach Number Theory to All Students* 203

COMMENTS: This activity is packed with computational practice. The final generalization can be closely related to triangular numbers. Be certain that all students manipulate the poker chips and compute the sums of the whole numbers as they look for a pattern.

Unlike Square Numbers (Activity 9.06), the initial pentagonal arrangement may be more difficult to make. The chips should be placed so they would form a regular pentagon if the appropriate tangent was drawn for each of them.

As the second pentagonal number is formed, the poker chips in that "layer" will have small spaces between them. The student should place a chip tangent to the existing chips for each side and then put the "corner" chips on.

Each student may need several chips so they will need to share or do this in small groups at a time.

TEACHER COMMENTS TO STUDENTS:

Take five poker chips from your pile. Make a regular pentagon with these five chips.

Now take 15 chips from your pile and make a regular pentagon. Did your pentagon look like this? (See Figure 9-14.)

Try to make a pentagon with 30 chips.

Try to make a pentagon with 45 chips.

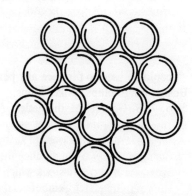

**FIGURE 9-14**

Use the chips and your answers above to fill in the chart shown in Figure 9-15.

Pentagonal Numbers

| Pentagonal Number? | 1 | 2 | 3 | 4 | 5 | 6 | 7 | 8 | 9 | 10 | 11 | 12 | 13 | 14 | 15 | 16 | 17 | 18 | 19 | 20 | 21 | 22 |
|---|---|---|---|---|---|---|---|---|---|---|---|---|---|---|---|---|---|---|---|---|---|---|
| | N | N | N | N | Y | N | N | N | N | N | N | N | N | N | Y | | | | | | | |

| 23 | 24 | 25 | 26 | 27 | 28 | 29 | 30 | 31 | 32 | 33 | 34 | 35 | 36 | 37 | 38 | 39 | 40 | 41 | 42 | 43 | 44 | 45 | 46 | 47 | 48 | 49 | 50 |
|---|---|---|---|---|---|---|---|---|---|---|---|---|---|---|---|---|---|---|---|---|---|---|---|---|---|---|---|
| | | | | | | | | | | | | | | | | | | | | | | | | | | | |

**FIGURE 9-15**

What will be the next pentagonal number after 50?
What patterns do you see in the chart?
Look at this pattern!

$5(1) = 5$
$5(1 + 2) = 5(3) = 15$
$5(1 + 2 + 3) = 5(6) = 30$
$5(1 + 2 + 3 + 4) = 5(10) = 50$
$5(1 + 2 + 3 + 4 + 5) = \underline{\qquad} = \underline{\qquad}$
$\underline{\qquad} = \underline{\qquad} = \underline{\qquad}$

Write a sentence which describes the above pattern.

Write a formula which you could use to get any pentagonal number.

### TITLE: Magic Square Cubes (9.08)

GOAL: Magic squares are familiar motivational devices for drill in addition, but many students become frustrated trying to create magic squares abstractly. To encourage reluctant students to experiment on an abstract level, this activity provides a concrete device to get them started.

| | |
|---|---|
| OBJECTIVE: | The student will create a magic square. |
| MATERIALS: | 9 cubes per student, a mirror. |
| GRADES: | 5, 6, 7, 8, General Mathematics (see Comments for Algebra I and Algebra II extensions). |
| INSTRUCTIONS FOR TEACHERS: | Many students cannot determine an immediate solution to a problem and refuse to experiment abstractly to find a solution. However, in working with cubes to make magic squares this difficulty may be overcome as students manipulate the cubes. |
| COMMENTS: | This activity is designed for third order magic squares, and consecutive whole numbers. You may wish to add more cubes to consider fourth and fifth order magic squares, or you may want to make magic squares with non-consecutive whole numbers, fractions, or integers. |
| | For algebra students refer to Ralph Manager's "An Algebraic Treatment of Magic Squares," *The Mathematics Teacher*, 1966, pp. 101-107, and to Henry Van Engen's "A Note on an Algebraic Treatment of Magic Squares," *The Mathematics Teacher*, 1966, p. 747. |

TEACHER COMMENTS TO STUDENTS:

A square made of nine cells in which the numbers in the rows, columns, and main diagonals have the same sum is called a magic square. Let's try to make one!

Get nine cubes. Write the numeral 1 on one face of six different cubes. Do the same for the numerals 2, 3, 4, 5, 6, 7, 8, and 9. When you have finished, all faces of the cubes should have a numeral written on them.

Try some arrangements of your cubes as shown in Figure 9-16. Is this a magic square?

When you make a magic square with your cubes, record your answer here.

|   |   |   |
|---|---|---|
| 9 | 1 | 7 |
| 5 | 6 | 3 |
| 8 | 4 | 2 |

**FIGURE 9-16**

Look at your answer carefully. Do you see another way to arrange the numbers to make a magic square? Put your mirror on the middle column and look at the reflection of your magic square. Is it still a magic square?_____ Is it a different magic square?_____Use your cubes and mirror to make different magic squares.